**Rescuing Eve is Rescuing Me - Adult Coloring Book**



**Published by "Write It Down Make It Plain"**

**Seattle, Washington**

**ISBN: 978-1-933133-60-7**



**Printed in the United States of America**

# Rescuing Eve Is Rescuing Me

## coloring journal

Written by
Naomi Peete

Illustrated by
Naomi Peete

REDEEM
SET FREE
RANSOM
DELIVERANCE
SAVE SOMEONE
LIBERATE
RESCUED

# Journal Notes

# This coloring journal belongs to:

# Dedicated to All the Daughters of Eve

APPRECIATION TO MICHELLE IHENACHO & GAIL HARRISON FOR THEIR POSITIVITY AND SUPPORT

The beat of a butterfly's wings in the west can cause a windstorm in the east.
Rescuing Eve and Me

# Content

EVE ME FAITH SCRIPTURE SCIENCE RESCUE IMAGE WHY? CO-DOMINION HUMANITY HER-STORY DIFFUSION BIOLOGY ABIOTIC BIOTIC STARDUST ELEMENTS CELL-THEORY CELLS LIKENESS MEDICAL-TECHNOLOGY ANESTHESIOLOGY.

# Journal Notes

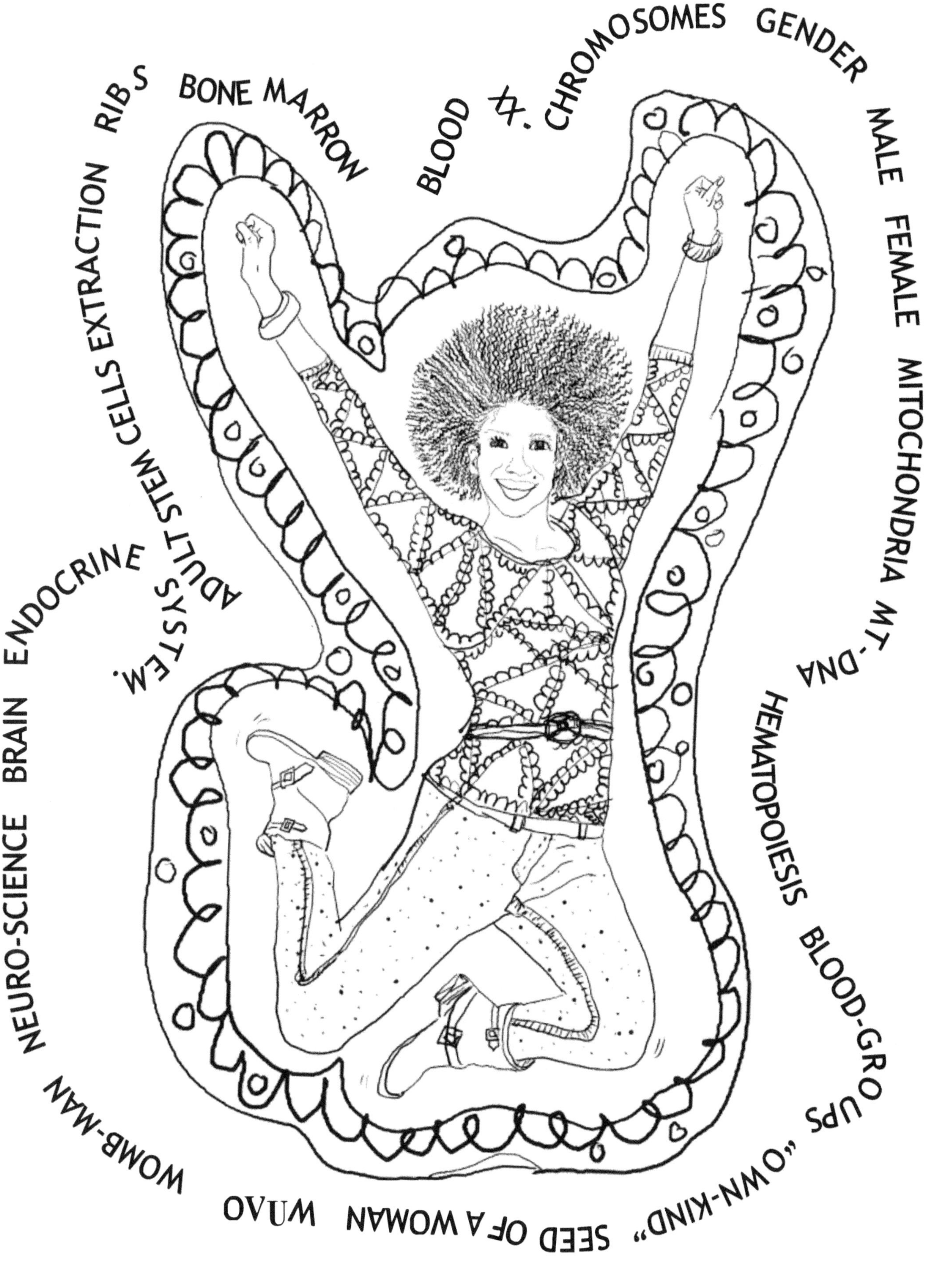
RIBS
BONE MARROW
BLOOD
XX. CHROMOSOMES
GENDER
MALE
FEMALE
MITOCHONDRIA
MT-DNA
HEMATOPOIESIS
BLOOD-GROUPS
"OWN-KIND"
SEED OF A WOMAN
OVUM
WOMB-MAN
NEURO-SCIENCE
BRAIN
ENDOCRINE SYSTEM.
ADULT STEM CELLS EXTRACTION

# Journal Notes

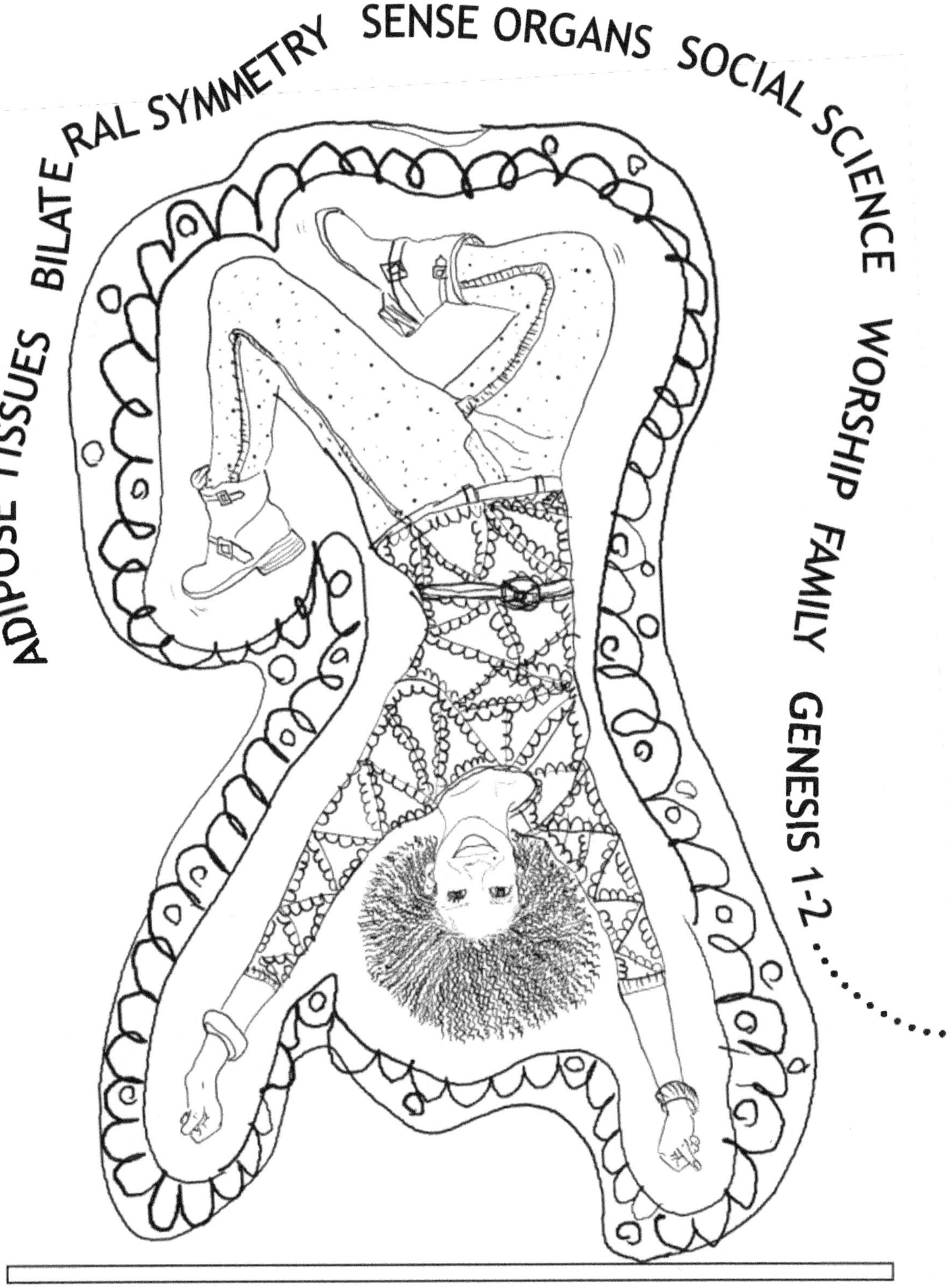
ADIPOSE TISSUES
BILATERAL SYMMETRY
SENSE ORGANS
SOCIAL SCIENCE
WORSHIP
FAMILY
GENESIS 1-2.......

# Journal Notes

# Pictures to Color of Women from all over the World

Nigeria. Japan. Brazil. Spain. Italy. Israel. South Africa. Botswana. Philippines. China. India. Pakistan, Thailand. Egypt. Korea. France. Germany. Mexico. Native American. Kenya. Dubai, Palestine, Greece, Australia, Ethiopia, Bahamas, United States, and more.

# Coloring Journal

**DO NOT BE ANXIOUS. COLORING AND JOURNALING ARE NOT WASTING TIME. IT IS A TIME FOR REFRESHING YOUR STRENGTH AND WISDOM. PHILIPPIANS 4: 6**

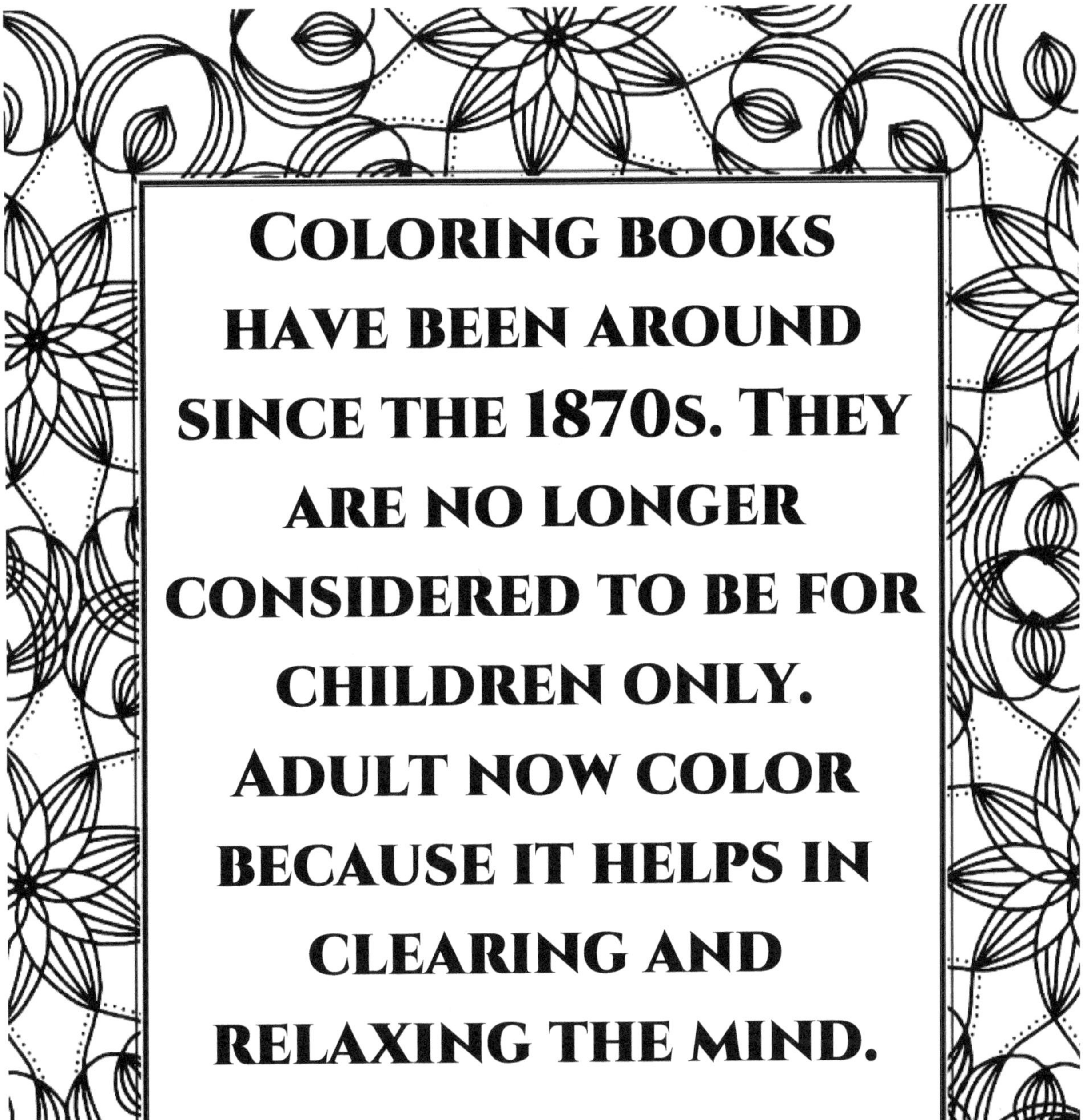

COLORING BOOKS HAVE BEEN AROUND SINCE THE 1870S. THEY ARE NO LONGER CONSIDERED TO BE FOR CHILDREN ONLY. ADULT NOW COLOR BECAUSE IT HELPS IN CLEARING AND RELAXING THE MIND.

# Benefits of Coloring Books

1. REDUCE STRESS AND ANXIETY – COLORING CAN RELAX THE FEAR CENTER OF YOUR BRAIN. IT INDUCES THE SAME STATE AS MEDITATING BY REDUCING THE THOUGHTS OF A RESTLESS MIND.
2. IMPROVE MOTOR SKILLS AND VISION – IT REQUIRES THE TWO HEMISPHERES (RIGHT AND LEFT) OF THE BRAIN TO COMMUNICATE. LOGIC HELPS US STAY INSIDE THE LINES, CHOOSING COLORS GENERATES A CREATIVE THOUGH PROCESS.
3. IMPROVE SLEEP – SCIENCE STATES YOU GET A BETTER NIGHT'S SLEEP WHEN AVOIDING ELECTRONICS AT NIGHT. EXPOSURE TO THE EMITTED LIGHT REDUCES YOUR LEVELS OF THE SLEEP HORMONE, MELATONIN. COLORING IS A RELAXING AND ELECTRONIC-FREE BEDTIME RITUAL THAT WILL NOT DISTURB YOUR LEVEL OF MELATONIN.
4. IMPROVE YOUR FOCUS – COLORING REQUIRES YOU TO FOCUS AND NOT BE STRESSFUL. IT ALLOWS YOU TO PUT EVERYTHING ASIDE AND LIVE IN THE MOMENT.

Color became
visible with light
(Genesis 1:3) and
it can be
separated out of
white light.
Genesis 9:13

# How to use Coloring Pencils

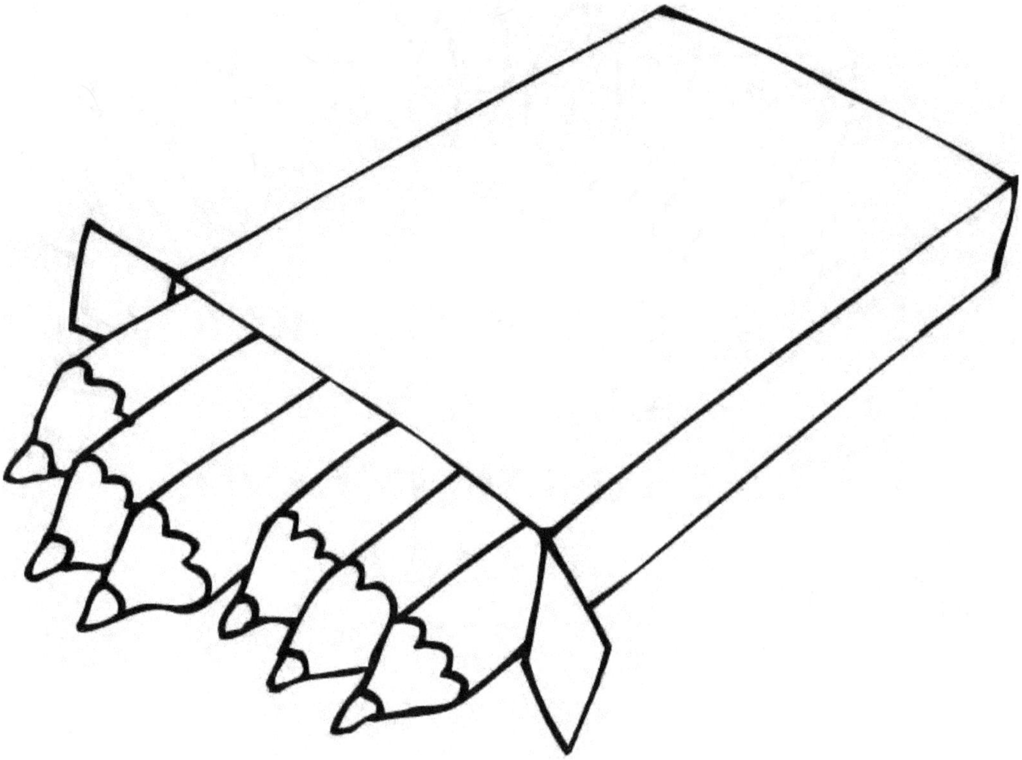

**COLORING PENCILS ARE GREAT TO USE BECAUSE**

- **THEY ARE INEXPENSIVE,**
- **HAVE A VARIETY OF COLORS,**
- **EASY TO USE,**
- **TRANSPARENT COLORS CAN BE LAYERED OVER EACH OTHER AND USING MULTIPLE LAYERS OF COLORS ON TOP OF ONE ANOTHER TO SHOW TEXTURE, SHADE, AND FORM.**

# Taxonomy – naming system Genesis 2:19

# Basic Colored Pencil Techniques

| Stippling | Hatching | Scumbling | Back and forth stroke | Cross-hatching |
|---|---|---|---|---|
| PLACING LOTS OF TINY DOTS ON YOUR IMAGE, CLOSE TOGETHER OR FAR APART- ADD TEXTURE | DRAWING PARALLEL LINES THAT ARE CLOSE AND REST ON TOP OF ONE ANOTHER | MOVING TIP OF PENCIL AROUND AN AREA IN UNEVEN CIRCLE | DRAW IN A CONTINUOUS BACK AND FORTH MOTION, WITHOUT LIFTING YOUR PENCIL OFF THE PAPER | DRAWING PARALLEL LINES AND DRAWING ANOTHER SERIES OF PARALLEL LINE IN ANOTHER DIRECTION ON TOP OF THE FIRST |
| 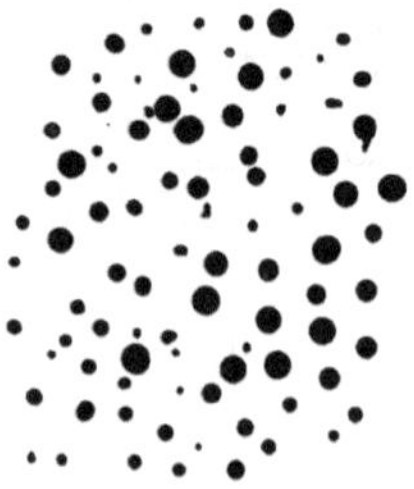 |  |  |  |  |

# Best way to sharpen colored pencils to prevent breakage —

1. THE LEAD IN YOUR COLORING PENCILS IS WAX OR OIL BASE. YOUR PENCILS WILL LAST LONGER IF THEY ARE SHARPENED WELL AND CARE FOR CORRECTLY.
2. YOU CAN USE HANDHELD PENCIL SHARPENERS, ELECTRIC PENCIL SHARPENERS, SANDPAPER, OR EVEN A KNIFE.
3. THE BEST WAY TO SHARPEN YOUR PENCIL IS TO HOLD THE PENCIL IN ONE HAND AND THE PENCIL SHARPENER IN THE OTHER HAND. THEN TURN THE PENCIL SHARPENER, WHILE HOLDING THE COLORED PENCIL STILL.
4. YOUR SHARPENER IS BEST WHEN YOU PRODUCE ONE LONG CONTINUOUS PENCIL SHAVING.
5. BROKEN BITS AND PIECES OF SHAVING INDICATES A DULL SHARPENER BLADE.

# GOD KNOWS MY NAME

**O Lord, You have searched me and known me. You know when I sit and when I rise; You understand my thoughts from afar (Psalm 139:1-2)**

As for Me and My House, We will Serve the Lord (Joshua 24:4:15).

# Science of Colors

# Journal Notes

# The Electromagnetic Spectrum

RADIO WAVES

MICROWAVES

INFRARED

VISIBLE LIGHT

ULTRAVIOLET

X-RAYS

GAMMA RAYS

ROY G. BIV

Visible Light separates into Colors

red

orange

yellow

green

blue

indigo

violet

"Let there be Light"

Genesis 1:3

# Color Wheel

**Primary Colors**

1. Red
2. Yellow
3. Blue

**Secondary Colors**

4. Green
5. Orange
6. Violet

**Tertiary Colors**

7. yellow-green
8. blue-green
9. blue-violet

10. red-violet
11. red-orange
12. yellow-orange

# Lydia

**If you judge me to be faithful to the Lord, come and stay at my house (Acts 16:15).**

In the Beginning God
Created the Heavens
and the Earth.
Genesis 1:1

# And God made Stars

## Genesis 1: 16

**LIGHTS IN THE FIRMAMENT OF THE HEAVEN TO DIVIDE THE DAY FROM THE NIGHT AND LET THEM BE FOR SIGNS, SEASONS, FOR DAYS, AND YEARS.**

**(GENESIS 1:14)**

# Rescuing Eve Is Rescuing Me

*I will praise You, for I am fearfully and wonderfully made after the likeness of my creator. Marvelous are Your works, and that my soul knows very well.*
*Psalm 139:13-14*

# Rescue Eve

# Why?

# Rescuing Eve is Rescuing Me

**To Reveal the Glory of God's original created template for womanhood through Eve.**

# To Rescue my:

- Image
- Identity
- Purpose
- Sanctity
- Thinking capacity
- Global
- .........

**To Enhance our under-Standing & appreciation of Eve, The first ezer, through the lens of science**

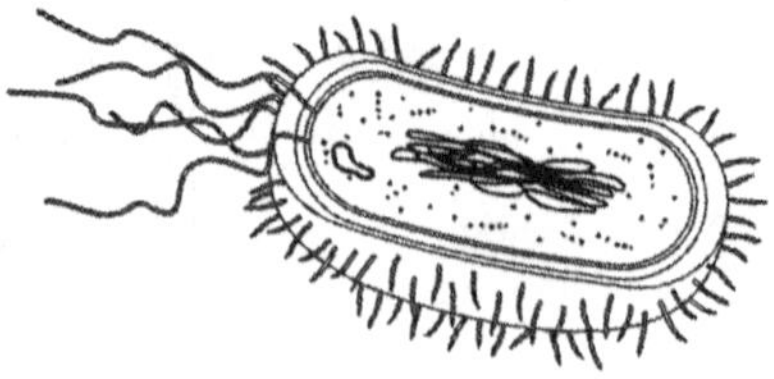

# Women of the World

Every woman comes with strengths and weaknesses. Down through the centuries the image of Eve is labeled with negativity. Eve was created in the image and likeness of her creator, Elohim. It is time to explore the uniqueness and significance of womanhood through scripture and science.

Discover the positive side of Eve. There is not much written about her in the scriptures. However, there is much we can learn from science and background events that will reposition all women to value their greatness as a daughter of Eve.

Search me, O God, and know my heart; test me and know my concerns. (Psalm 139:24)

# Co-Dominion

**Who was Eve? Eve is the first woman, first wife, the first mother, and grandmother in the Bible. She shared dominion over the fish of the sea, and over the fowl of the air, over the cattle, over the creeping things, and over all the earth.**

**Genesis 1:26**

# Humanity Established

**EVE WAS ESSENTIAL IN CREATING THE BASIC UNIT OF SOCIETY CALLED THE FAMILY. AS HUMANITY INCREASED FAMILY HISTORY WAS FORMED WHICH LEAD TO THE DEVELOPMENT OF CIVILIZATION (GENESIS 4, 5).**

# My Soul Thirst for God

## Psalm 42:2

MARY MAGALENE-"
loved by Yahweh"

# Her-Story

**Eve was articulate, attractive, anointed, appointed, curious (Genesis 3:2), an independent thinker, shortsighted at times, industrious, and a worshipper (Genesis 4:25). A shared decision of disobedience (Genesis 3:6) that caused a dramatic shift in her history and the history of the world.**

Diffusion - movement of molecules in the air

**Three Women who anointed Jesus: Luke 7:36-50; John 12:1-8; Matthew 26:1-5, Mark 14:3**

# Journal Notes

Rebekah - Nations are in your womb - Genesis 25:23

# The Shunammite Woman

I MUST GO QUICKLY TO THE MAN OF GOD. DRIVE ONWARD; DO NOT SLOW THE PACE. SO, SHE SET OUT AND WENT TO SEE ELISHA AT MOUNT CARMEL. – 2 KINGS 4: 18-27

# Journal Notes

# Deborah- "the Bee", Prophetess, fourth Judge of Israel, Wife, and Heroine - Judge 4, 5

# Journal Notes

The first couple was told to subdue and have dominion over all the creatures of the air, water, and land. To accomplish this task, they had to observe patterns, study behaviors, structures, diets, habitat, and life cycles. Genesis 1:26-28

**ABIOTIC (NONLIVING) FACTORS GENERATED THE MATERIALS FOR THE CLAY MAN: SOLAR ENERGY, THE ATMOSPHERE, TOPOGRAPHY, OXYGEN, AND MINERALS AND WATER. THESE SUBSTANCES GENERATED ADAM.**

**BIOTIC (LIVING) FACTORS ARE THE LIVING THINGS THAT SHAPE THE ENVIRONMENT (ANIMALS, PLANTS, BACTERIA, OTHER EXISTING CELLS). ALL THESE FACTORS FUNNEL DOWN TO THE NEW BEGINNINGS IN CREATING EVE.**

# Dust of the Earth and Water

Genesis 2:6-7 - The Clay man created from water and red dirt.

# Journal Notes

# Abiotic Materials

**ADAM IS THE ONLY HUMAN ON RECORD THAT WAS PRODUCED FROM NON-LIVING MATTER OR INANIMATE MATERIALS. ADAM WAS CREATED FROM THE DUST OF THE EARTH. ALL LIFE AFTER ADAM ORIGINATED FROM CELL. THE SMALLEST UNIT OF LIFE.**

# Stardust

SCIENCE EXPLAINS THAT THE HUMAN COMPOSITION CONSISTS OF SEVERAL BASIC ELEMENTS. BY MASS 99% OF THE HUMAN BODY IS MADE UP OF SIX ELEMENTS OXYGEN, CARBON, HYDROGEN, NITROGEN, PHOSPHORUS, AND SULFUR. THESE ELEMENTS EXIST IN OUR GALAXY AND BEYOND. WE ARE STARDUST.

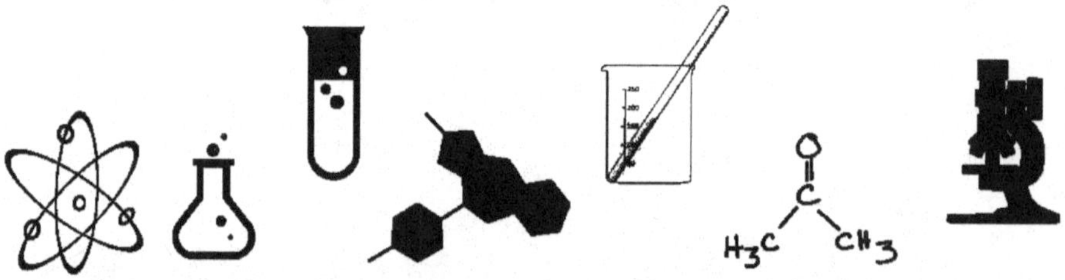

## Essential Elements in Composition of the Human Body

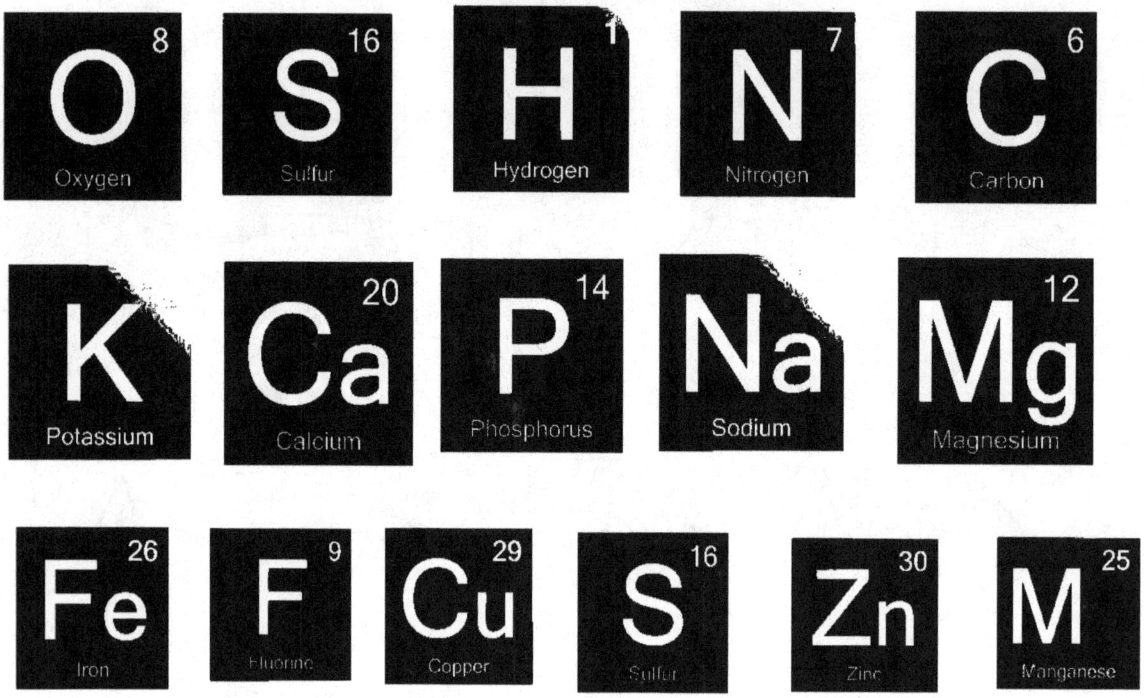


**IN THE BEGINNING GOD USED SMALL SIZES AND QUANTITY AND LIMITED THE NUMBER OF COMPONENTS TO CREATE COMPLEX STRUCTURES THAT RESTED ON THE FOUNDATION OF A FEW ESSENTIAL ELEMENTS.**

- **Oxygen**- 65% OF YOUR BODY'S WEIGHT; AEROBIC RESPIRATION USED IN MITOCHONDRIA TO MAKE ATP (ENERGY); FOUND IN ORGANIC MOLECULES; LARGE AMOUNT IN LUNGS AND BLOODSTREAM; BINDS TO HEMOGLOBIN IN RED BLOOD CELLS; COMBINES WITH HYDROGEN TO MAKE WATER
- **Sulfur** – FOUND IN AMINO ACIDS TO BUILD PROTEINS
- **Hydrogen**- 10% OF BODY WEIGHT; INVOLVED IN CRITICAL CHEMICAL REACTIONS OF FORMING AND BREAKING CHEMICAL BONDS; BINDS WITH OXYGEN TO FORM WATER; FOUND IN ALL ORGANIC MOLECULES
- **Nitrogen**- 3% OF BODY WEIGHT; OBTAIN FROM EATING PLANTS OR HERBIVORES; NEEDED IN PROTEIN SYNTHESIS FOR NUCLEIC ACIDS (DNA, RNA)

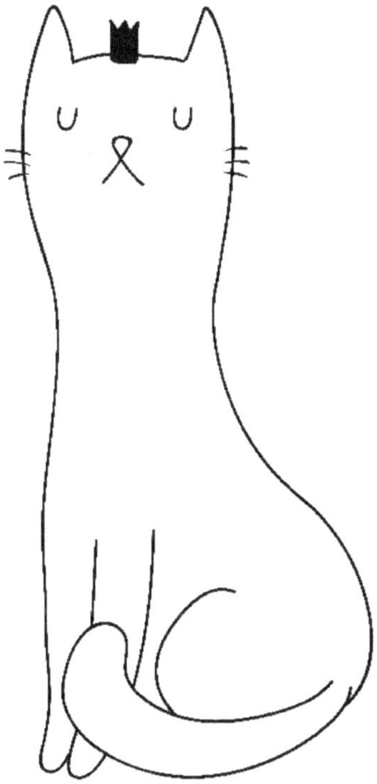

- **Carbon – 18% of body weight; makes up molecules base for organic compounds**
- **Potassium – forms electrolytes in blood for maintaining muscular and nervous systems; regulates the heartbeat**

- **Calcium- 1.4 % of body weight; mostly found in the bones and teeth; needed in muscular contraction and protein regulation**
- **Phosphorus-1% of body weight; found in bones, teeth, nucleic acids and energy molecules**
- **Sodium- essential in nerve transmission and muscle function**

- **Magnesium – ENZYME COFACTOR; BINDS TO ATP AND NUCLEOTIDES; USED TO BUILD HEALTHY TEETH AND BONES**
- **Iron--ENZYME COFACTOR; SYNTHESIS OF HEMOGLOBIN AND ELECTRON CARRIERS; COMPONENT OF BLOOD AND USED IN OXYGEN TRANSPORT**
- **Fluorine - MAINTENANCE OF TOOTH STRUCTURE**
- **Copper-IMPORTANT AS AN ELECTRON DONOR IN BIOLOGICAL REACTION; REGULATES IRON TRANSPORT**
- **Zinc-ESSENTIALS TRACE ELEMENT THAT HELPS TO REGULATE GENES**
- **Manganese-ESSENTIAL FOR CERTAIN ENZYMES THAT PROTECT THE MITOCHONDRIA**

**God supplies all our needs - Psalm 23**

# Journal Notes

# Cell Theory- Genesis 2:22

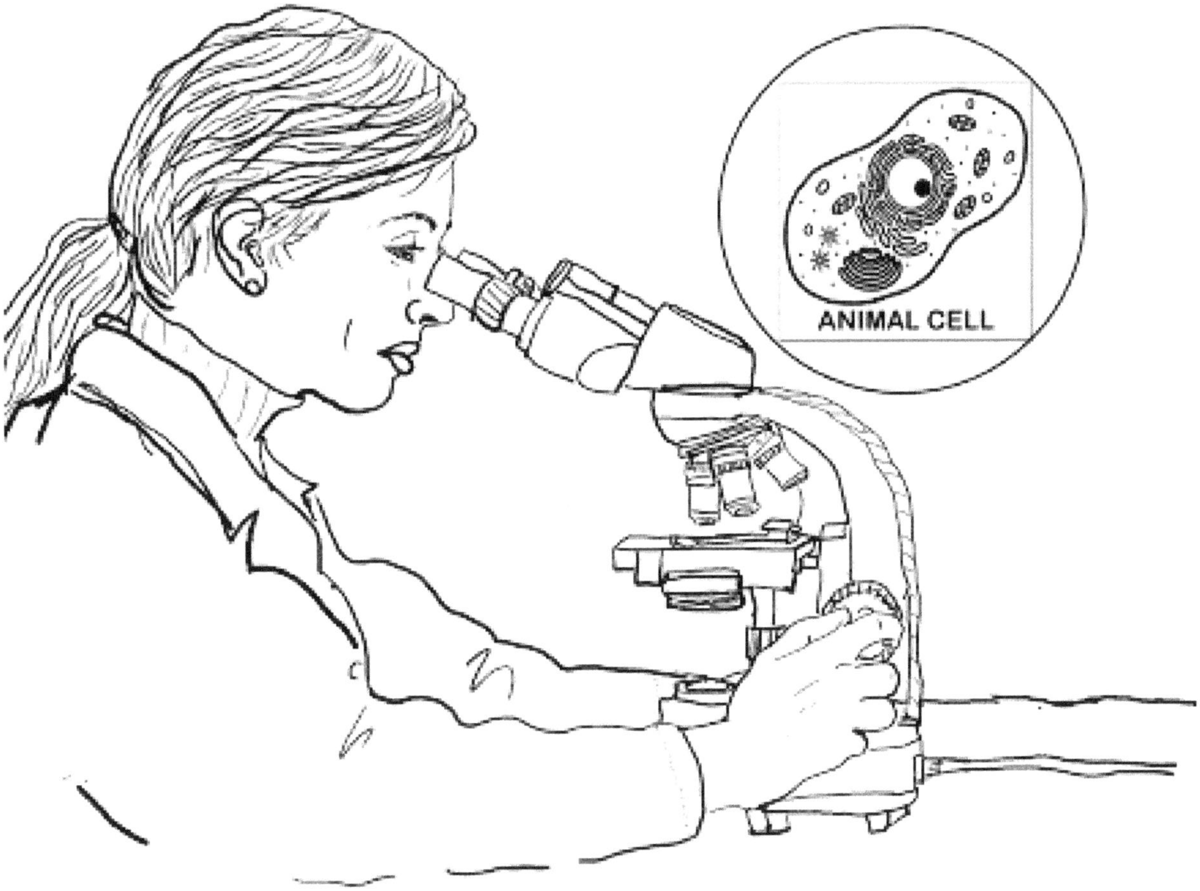


ALL LIVING THINGS ARE COMPOSED OF CELLS. CELLS ARE THE SIMPLEST UNIT OF LIFE. WE ARE COMPOSED OF BILLIONS OF CELLS GENERATED BY DIVISION CALLED MITOSIS. THREE GERMANY SCIENTISTS ARE CREDITED AS THE FOUNDERS OF THE CELL THEORY. UNDERSTANDING THE CELL THEORY EXPLAINS LIFE, GROWTH, AND DEATH.

# Journal Notes

**The main points of the Cell Theory are:**

- **All living things are made of cells**
- **Cells are the basic building blocks of all life**
- **All cells came from pre-existing cells generated through the process of cell division**
- **Energy flow occurs within and between cells**
- **Hereditary information is passed from cell to cell**
- **All cells have the same basic chemical composition: water, inorganic ions, and organic molecules**

# Journal Notes

The Cell Theory allows us to differentiate between what is living and non-living. Lifeforms today have descended in an unbroken continuity from the cells of Adam and Eve. Creation give us proof to the science of life (Genesis 1:1). Eve is the 'Mother of all Living'. The creation of Eve is the culmination of all

THE CREATION OF EVE SUPPORTS THE CELL THEORY THAT ALL CELLS COME FROM PRE-EXISTING CELLS. THE CELLS FOR THE CREATION OF EVE CAME FROM ADAM'S RIB CELLS.

CELLS"

- ARE THE BASIC SOURCE OF REPRODUCTION MATERIALS (DNA)
- PROVIDE STRUCTURE FOR THE BODY, TAKE IN NUTRIENTS FROM FOOD
- CARRY OUT ALL IMPORTANT FUNCTIONS ESSENTIAL TO LIFE (EXCRETION, ASSIMILATION, SYNTHESIS, DIGESTION)
- GROUPS OF CELLS THAT PERFORM SPECIFIC FUNCTIONS ARE CALLED TISSUES (NERVOUS, MUSCLES, BONE, BLOOD, SKIN)

# A Typical Cell

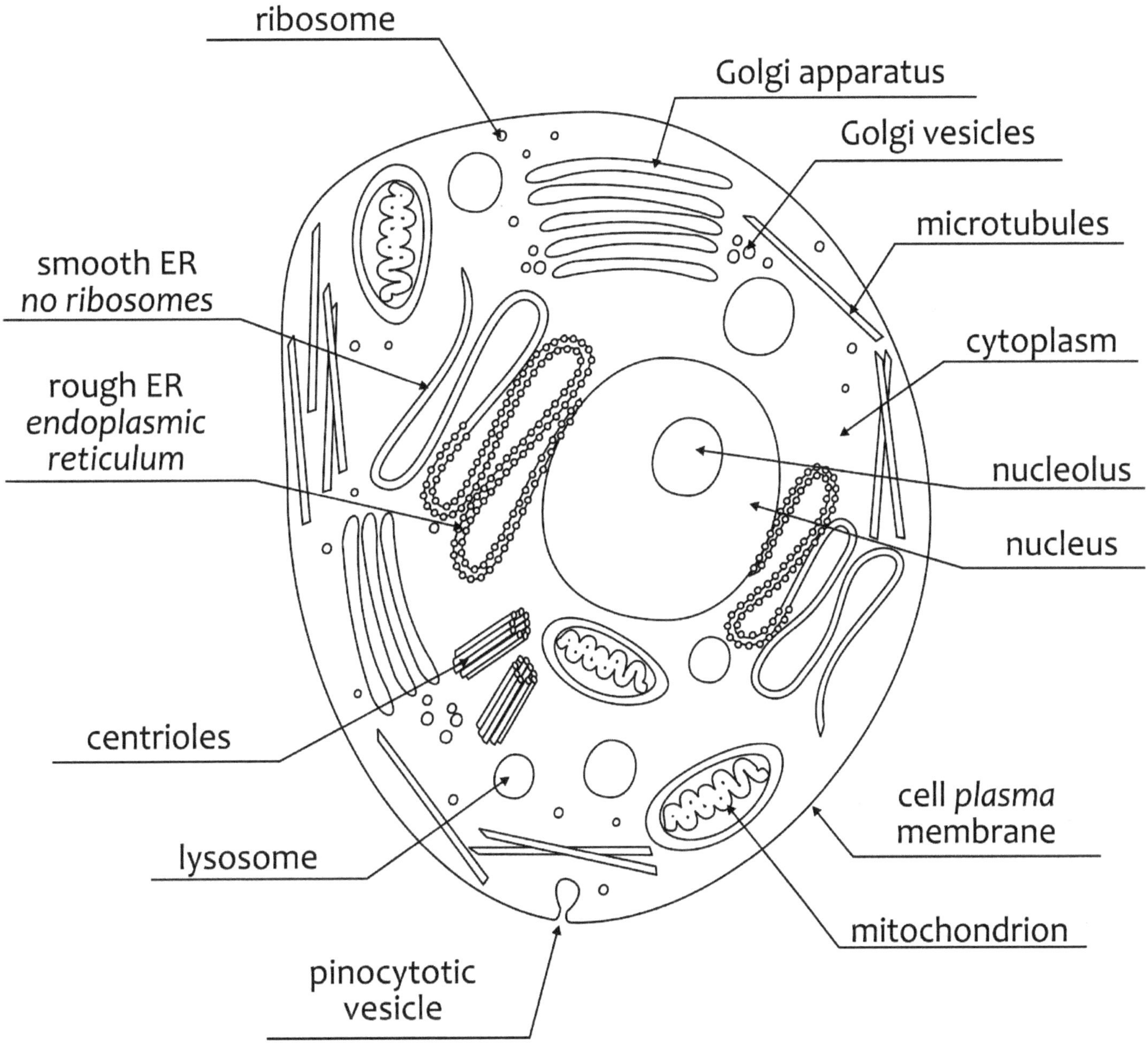

# Journal Notes

# Devout and Prominent Women

Anna
Sarah
Mary

# Journal Notes

And God said, Let us make woman in our image , after our likeness -

# Osteology – the study of Bones

**ADAM WAS PUT TO SLEEP AND GOD TOOK HIS RIBS AND CLOSED UP THE FLESH – GENESIS 2:21**

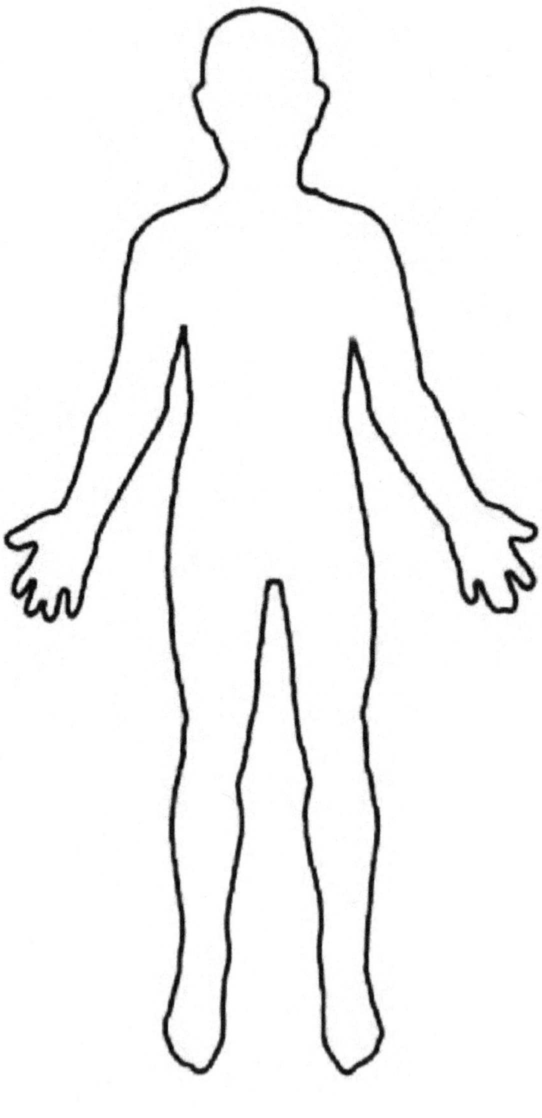

# Medical Technology

MEDICAL TECHNOLOGY: USE OF KNOWLEDGE AND MATERIALS TO BUILD AND IMPROVE PEOPLE'S LIVES.

TECHNOLOGY IS THE APPLICATION OF SCIENCE FOR PRACTICAL PURPOSES. OUR CREATOR DID NOT START WITH DUST TO FORM THE BEGINNINGS OF EVE. SPECIAL CELLS WERE EXTRACTED FROM BONE MARROW OF THE RIBS. BONE MARROW IS A RICH SOURCE OF MULTIPOTENT STEM CELLS THESE CELLS ARE THE PROGENITOR OR STEM CELLS STARTERS.

# Anesthesiology and Surgery

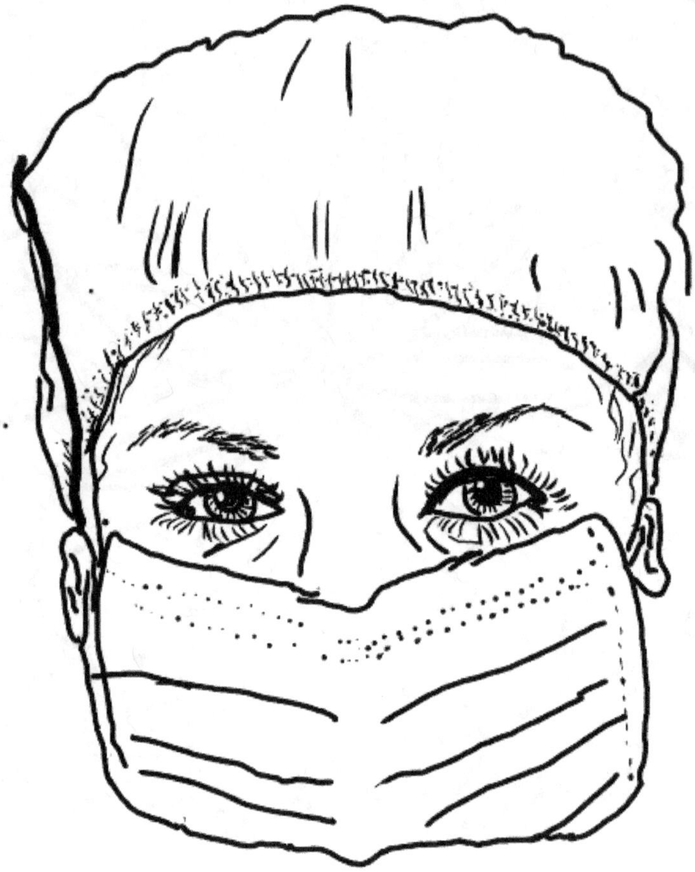

**SURGERY IS A MEDICAL PRACTICE THAT TREATS INJURIES, DISEASES, REPAIRS OR READJUST ORGANS AND TISSUES OFTEN SURGERY INVOLVES CUTTING INTO THE BODY. ANESTHESIS IS FOR SLEEEP AND PAIN REDUCTION. ADAM WAS THE WORLD'S FIRST PATIENT. HIS SURGERY PROVIDED THE MATERIALS NEED FOR THE BEGINNINGS OF EVE. GENESIS 2:21**

# Adult Stem Cells Extractions

THE PRESENT DAY MEDICAL PROCESS OF OBTAINING ADULT STEM CELLS FOR THERAPY IS VERY SIMILAR TO THAT OF ADAM'S RIB BONE EXTRACTION. THE MEDICAL TECHNOLOGIST WILL:

- DECIDE ON A DESIGNATED BONE EXTRACTION SITE
- PREPARE THE DONOR BY USINGLOCAL ANESTHETICS THAT IS ADMINSTED BY A CERTIFIED PHYSICIAN
- AN INCISION IS MADE
- STEM CELLS ARE EXTRACTED FROM THE BONE
- PATIENT IS SULTURED
- REST, RECOVERY, AND HEALING TIME IS NEEDED BEFORE THE PATIENT IS RELEASED

# The Rib- Genesis 2:21

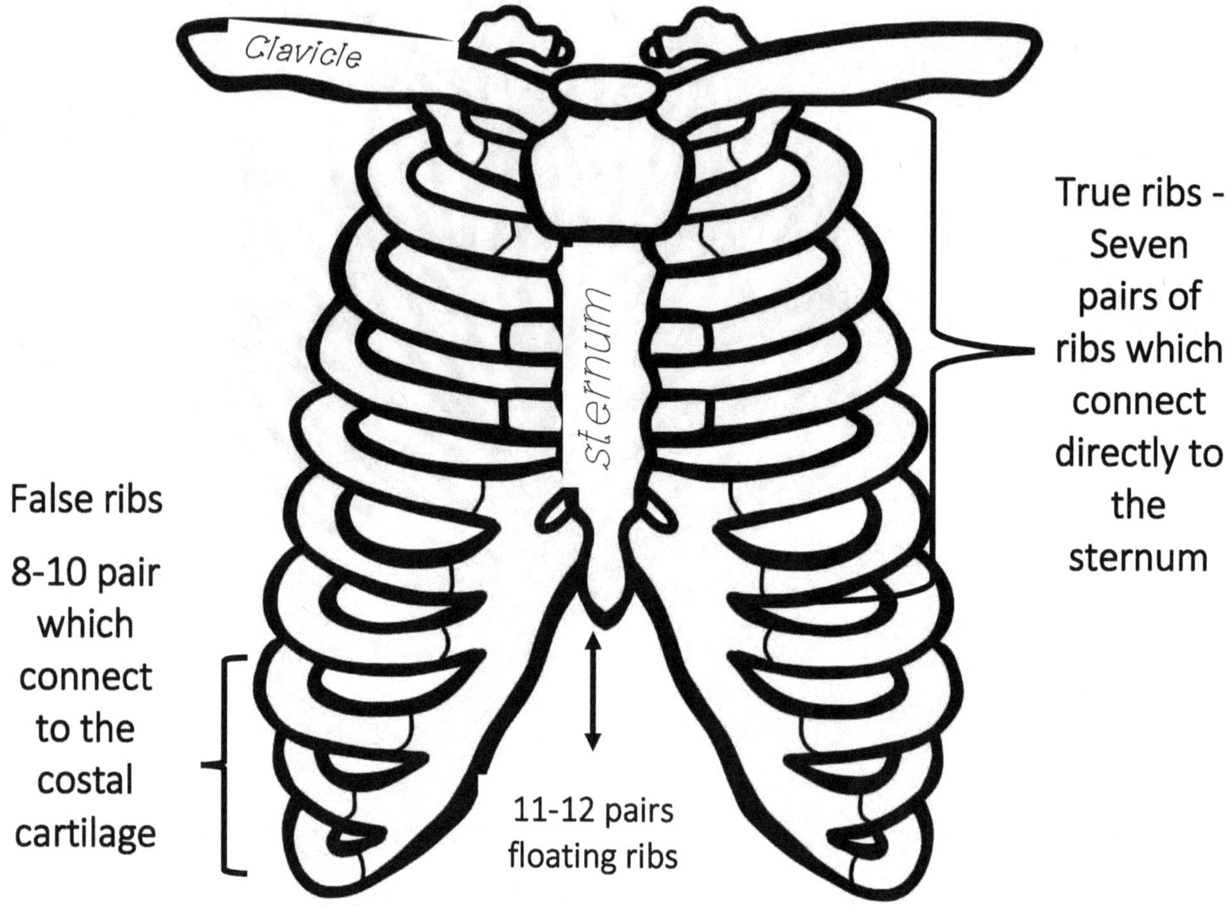


**THE RIB BONE IS ONE OF THE FEW BONES MENTIONS BY NAME IN THE SCRIPTURE. THE ESSENTIAL BUILDING UNIT OF THE FIRST WOMAN CAME OUT OF THE FIRST MAN. THE CREATION OF WOMAN WAS THE BEGINNING OF A CHRONOLOGICAL INCIDENT IN THE HISTORY OF HUMANS.**

**THE RIBS ARE A SPECIAL TYPE OF LONG CURVED BONES WITH A SPONGE LIKE INTERIOR THAT IS RICH IN BLOOD MAKING MATERIALS. RIBS PROVIDE SUPPORT AND PROTECTION FOR SOFT ORGANS AND ATTACHMENTS POINTS FOR CHEST MUSCLES.**

ALTHOUGH ALL BONES CAN REPAIR THEMSELVES, RIBS CAN REGENERATE THEMSELVES. RIBS ARE COMMONLY REMOVED DURING SURGERIES THAT REQUIRE BONE GRAFTS IN OTHER PARTS OF THE BODY. THE PERIOSTEUM (MEMBRANE AROUND THE BONE) MUST REMAIN, AS IT CONTAINS OSTEOBLAST) BONE MAKING CELLS) WHICH BUILD THE NEW RIB BONE. GOD CHOSE A MAJOR BONE THAT COULD REGENERATE ITSELF FROM THE BODY OF ADAM.

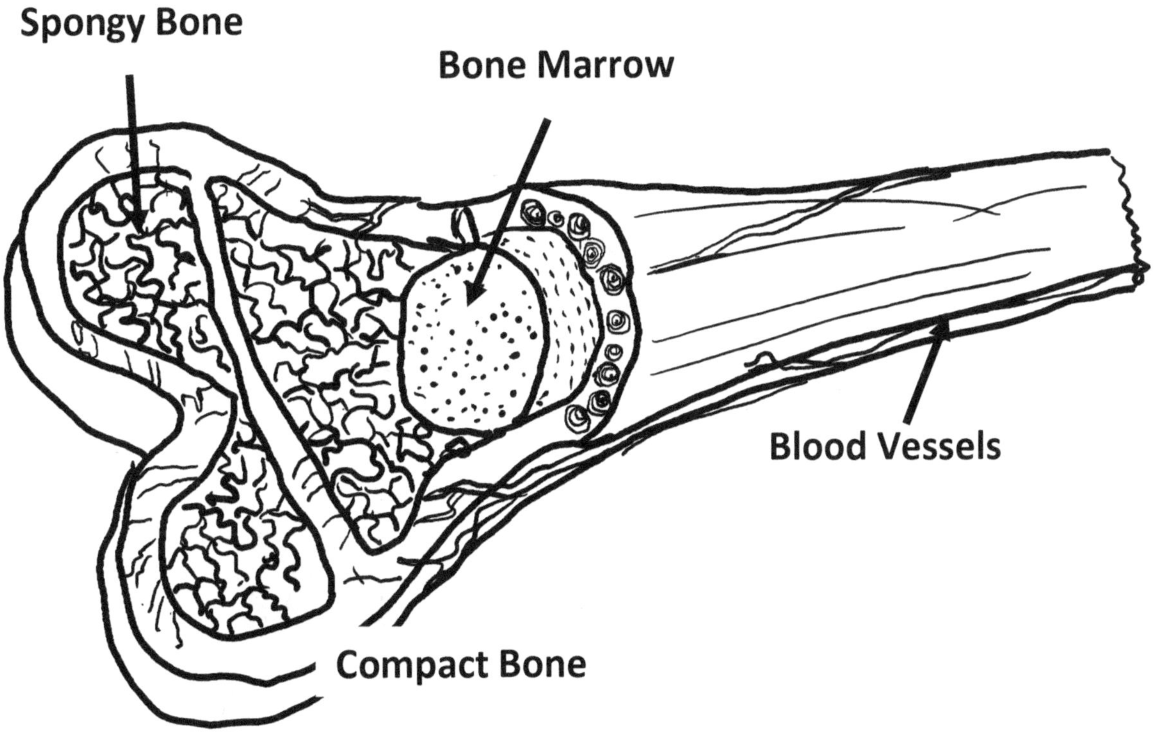


HUMAN RIBS CONTAIN BONE MARROW. OTHER LONG BONES CONTAINING BONE MARROW INCLUDE THE HIPBONES, SKULL, VERTEBRATE, BREASTBONE, AND SHOULDER BONES. BONE MARROW IS THE TISSUE INSIDE LARGE BONES AND WHERE STEM CELLS PRODUCE NEW BLOOD CELLS.

**HUMAN CANNOT REGENERATE LOST LIMBS, BUT THEY CAN REPAIR LARGE PORTIONS OF THEIR RIBS.**

**Always be joyful Never stop praying. Be thankful in all circumstances.**

**1 Thessalonians 5:16-18**

# Journal Notes

# EVE, a Creation from Stem Cells

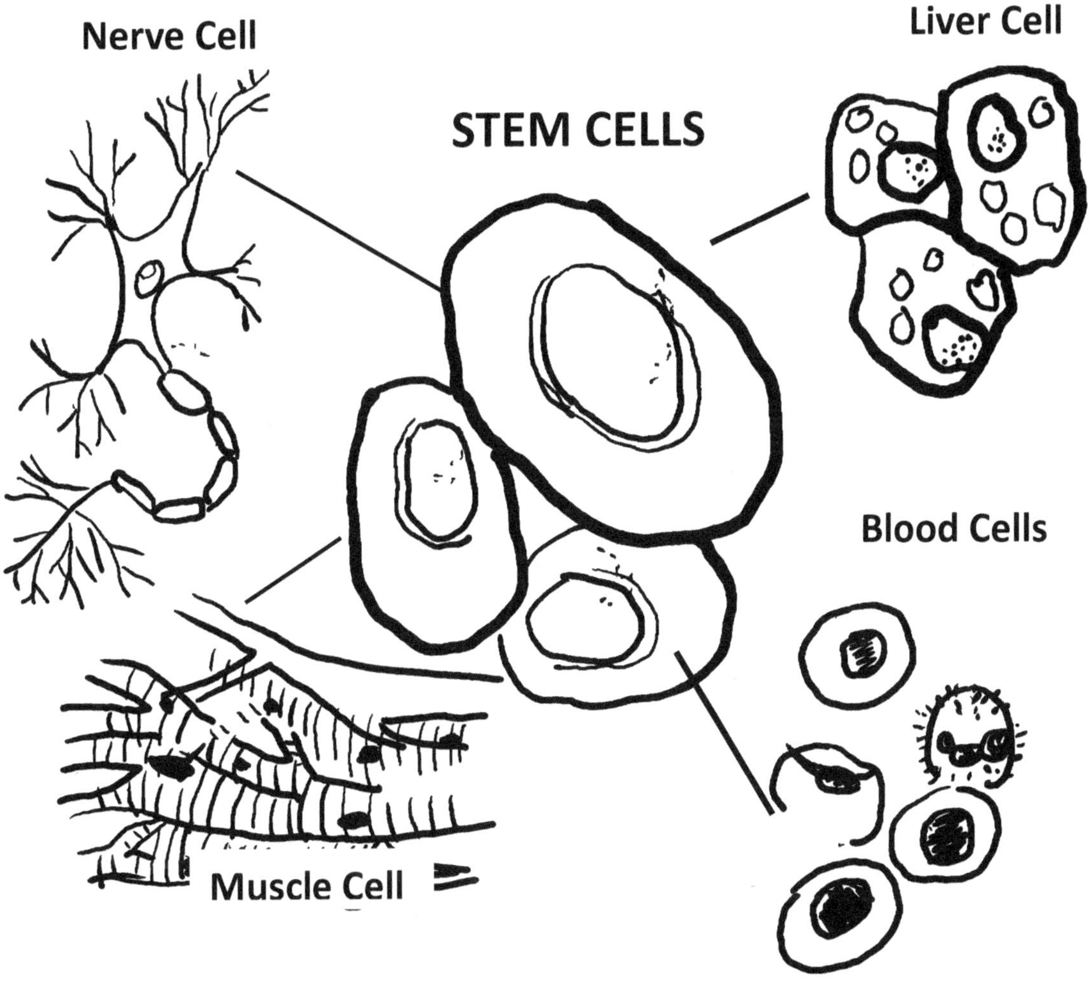


**THESE AMAZING CELLS EXTRACTED FROM BONE MARROW ARE CALLED STEM CELLS. STEM CELLS POSSESS MIRACULOUS SELF-RENEWAL POTENTIAL. STEMS CELLS HAVE THE POTENTIAL TO BUILD DIFFERENT TYPES OF CELLS.**

Stems cells are present both before birth (embryonic stem cells) and in the adult body (adult stem cells).

Eve was created from Adam's rib bone cells. These cells would be used to regenerate tissue and organs. The stem cells from Adam's rib bone marrow would undergo cell division called mitosis and develop into other organized cell structures.

Stem cells can develop into the four basic types of tissues: epithelia (provides an outer covering and boundary (the skin); nerve tissues (transmit electrical message from the brain to all body regions; muscles (contract and provide movement shape, heat transmission); and connective tissues (bind , support, and hold things together).

# Journal Notes

# TYPES OF CELLS IN YOUR BODY

STEMS CELLS ARE THE BUILDING BLOCKS OF THE HUMAN BODY. THESE CELLS DIVIDE OVER AND OVER AGAIN (MITOSIS) TO CREATE A FULL PERSON.

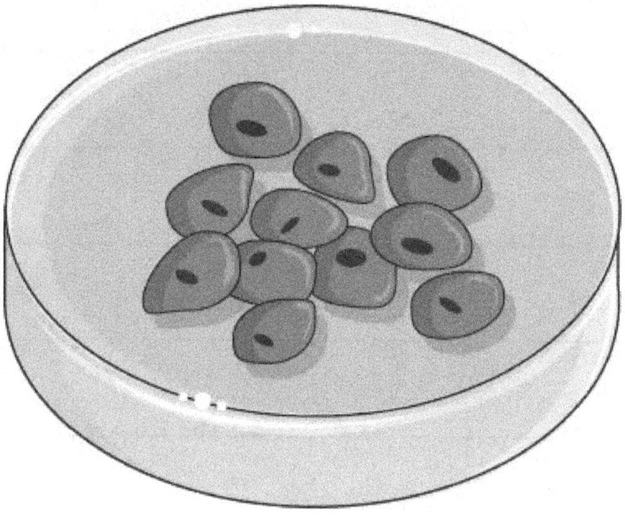

ADULT STEM CELLS FROM BONE MARROW CAN BECOME ANY TYPE TISSUE (GROUPS OF SIMILAR CELLS) IN THE BODY.

CELLS COME IN MANY SHAPES AND SIZES. EACH CELL IN THE BODY HAS A SPECIALIZED FUNCTION BASED ON TISSUE FORMATION. THE SHAPES OF CELLS PROVIDE CLUE TO THEIR FUNCTIONS.

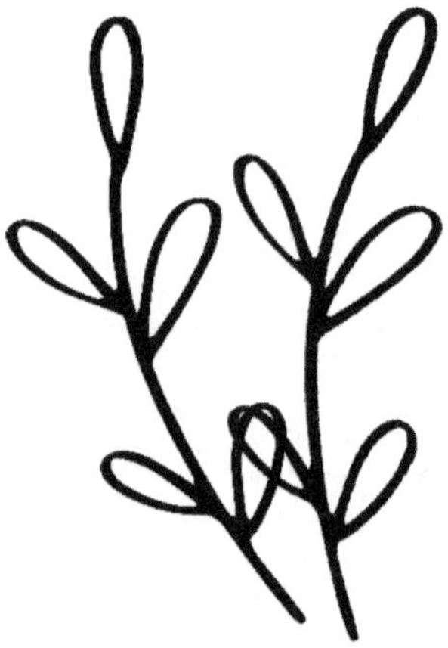

SOME CELLS ARE ELONGATED, TORPEDO SHAPED, ROUND, SLENDER AND TAPERED, FLAT AND LOCKED TOGETHER LIKE A PUZZLE, WHIP-LIKE TAILS, AND EVEN STARED SHAPED.

FUNCTIONS CAN VARY. CELLS CAN CARRY FLUIDS, STORE NUTRIENTS, TRANSPORT ELECTRIC SIGNALS, TRAVEL

THROUGH THE BODY, CHANGE SHAPES AND SQUEEZE INTO NARROW TUBES, AND SOME ARE FLEXIBLE.

# DIFFERENT TYPES OF CELLS

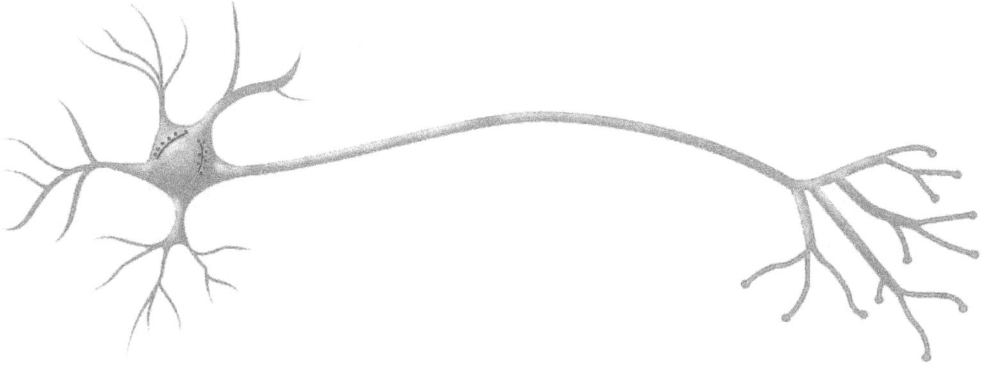

NERVE CELL – LONG. THIN, HAVE MANY BRANCHES, CARRIES ELECTRICAL MESSAGES FROM THE BRAIN TO THE REST OF THE BOY AND BACK, PRESENT ALL OVER THE BODY, AND THEY NEVER MULTIPLY OR DIVIDE IN ONE'S LIFETIME.

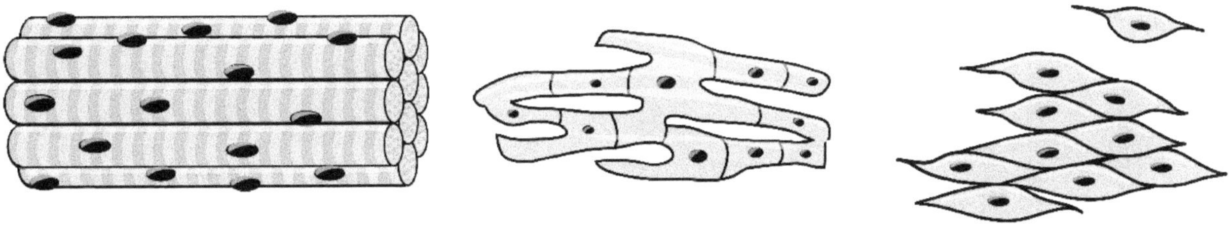

MUSCLE CELLS- LARGE, ELONGATED, SPINDLE-LIKE FIBERS. ALLOWS FOR CONTRACTIONS THAT PRODUCES MOVEMENT AND HEAT. THEY ARE THREE TYPES--SKELETAL, CARDIAC, AND SMOOTH MUSCLE CELLS

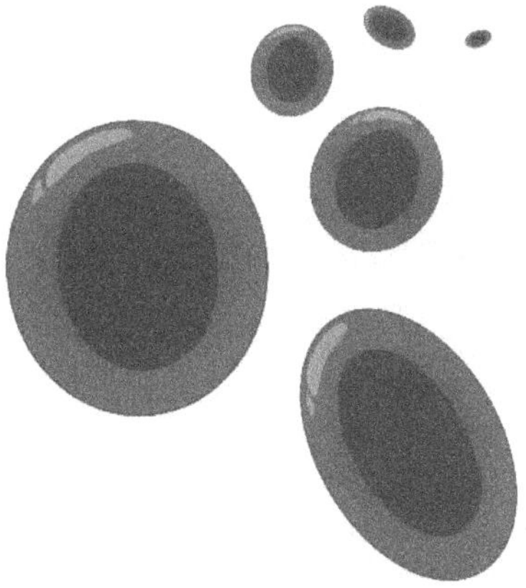

RED BLOOD CELLS -BICONCAVE SHAPE THAT ALLOWS FOR RAPID OXYGEN ABSORPTION, ALWAYS MOTILE AND NEVER STAY IN ONE PLACE, NEVER MULTIPLY, LIMITED LIFESPAN, AND BAG OF OXYGEN-CARRYING HEMOGLOBIN MOLECULES

**ADIPOSE CELL—BULKY, CRAMMED WITH DROPLETS OF FAT, ENERGY STORAGE SOURCE, AND REDUCES FRICTION IN THE BODY. CAN BE FOUND IN THE SOLES, PALMS, AND BABY CHEEKS.**

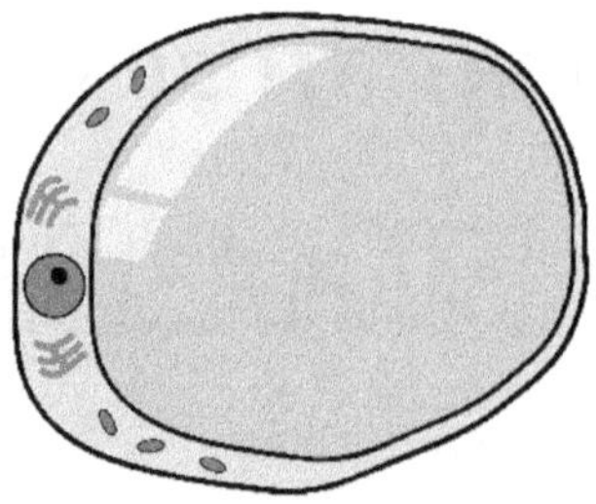

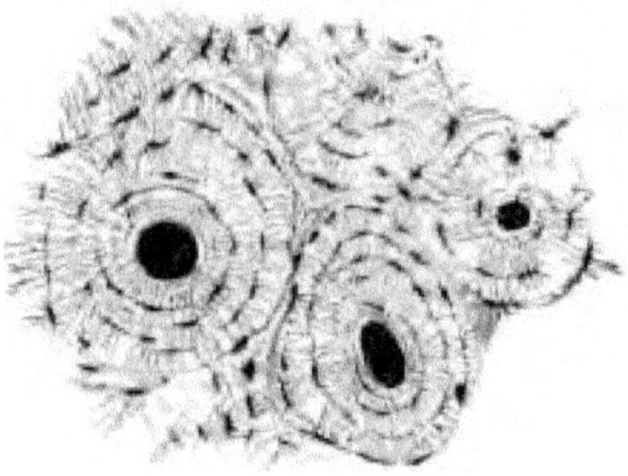

**BONE CELLS – MINERALIZED CONNECTIVE CELLS, SUPPORT, PROTECTIVE, STRENGTH**

**SPERM CELL—THE HEAD CARRIES THE PATERNAL MATERIALS, HAS WHIP-LIKE TAIL WHAT PROPELS IT TOWARDS THE EGG**

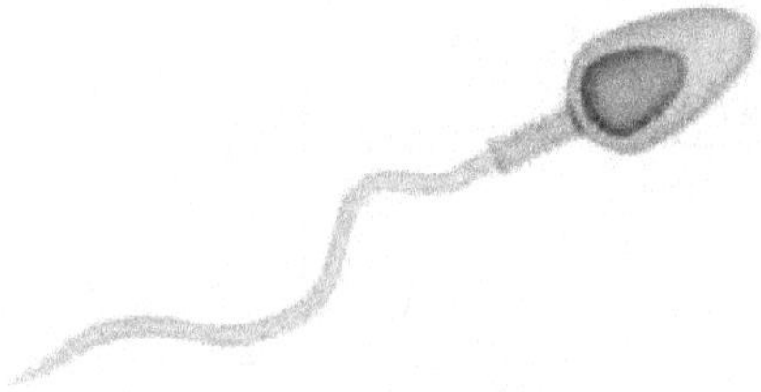

**PHOTORECEPTOR CELLS – ACTIVATED BY BRIGHT LIGHT AND RESPONSIBLE FOR COLOR PERCEPTION**

**EPITHELIAL CELLS –**

**SIMPLE, FLAT,**

**CELLS THAT COVER MOST ORGANS, LINES HOLLOW CAVITIES. FOUND IN THE SKIN RESPIRATORY TRACT, MOUTH CAVITY, AND SURFACE OF THE HEART**

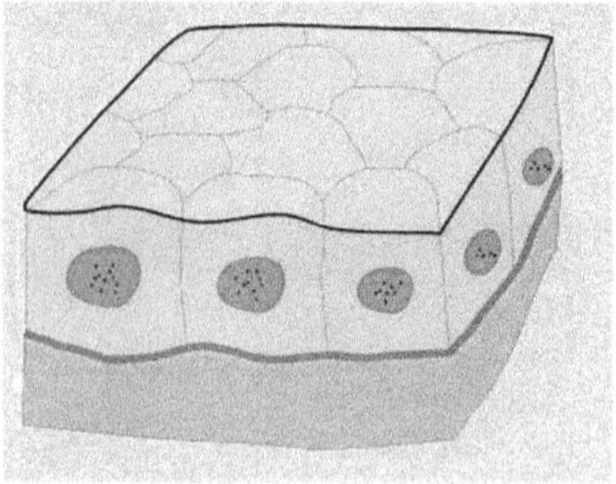

## Gender Specific Creation- Genesis 1:27

GENDER IS BIOLOGICAL DETERMINED BASED ON YOUR REPRODUCTIVE ORGANS AND FUNCTIONS. "AND PHARISEES CAME UP TO JESUS AND TESTED HIM BY ASKING, "IS IT LAWFUL TO DIVORCE ONE'S WIFE FOR ANY CAUSE?" HE ANSWERED, "HAVE YOU NOT READ THAT HE WHO CREATED THEM FROM THE BEGINNING MADE THEM MALE AND FEMALE?" MATTHEW 19:3-5.

# XX Chromosomes Extraction

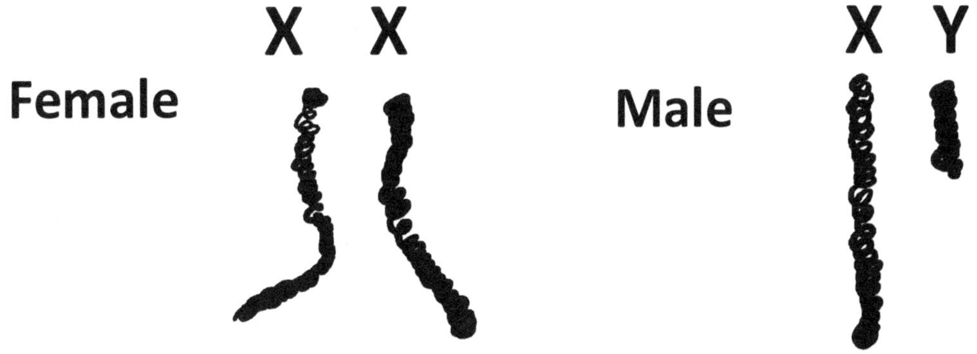


Adam and Eve are biologically exceptional creation. Although both shared similarities they both possessed distinct differences. These differences prepared them for their roles in creation.

Eve had a 'XX' pair of sex chromosomes, and Adam, a 'XY 'pair. A baby's gender is determined by the sperm cell that fertilizes a woman's egg. Sperm carries one sex chromosome, either a' Y' (male) or 'X' (female). The female 'X' chromosomes have more working genes that the smaller 'Y' chromosomes. The primary function of the 'Y' chromosomes is to give an organism its maleness.

# Women are the Carriers of History

# THERE IS A LITTLE BIT OF EVE IN ALL OF US

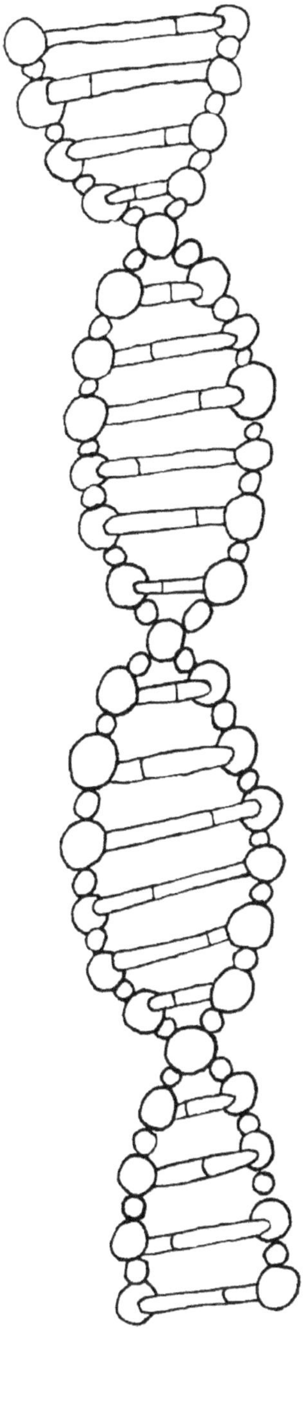

**A MITOCHONDRIAL DNA TEST TRACES A PERSON'S MOTHER LINE OF ANCESTRY USING THE DNA IN HIS OR HER MITOCHONDRIA. BASICALLY, MTDNA IS PASSED DOWN BY THE MOTHER UNCHANGED TO ALL HER CHILDREN, BOTH MALE AND FEMALE.**

# Mitochondria

## Powerhouse of the Cell

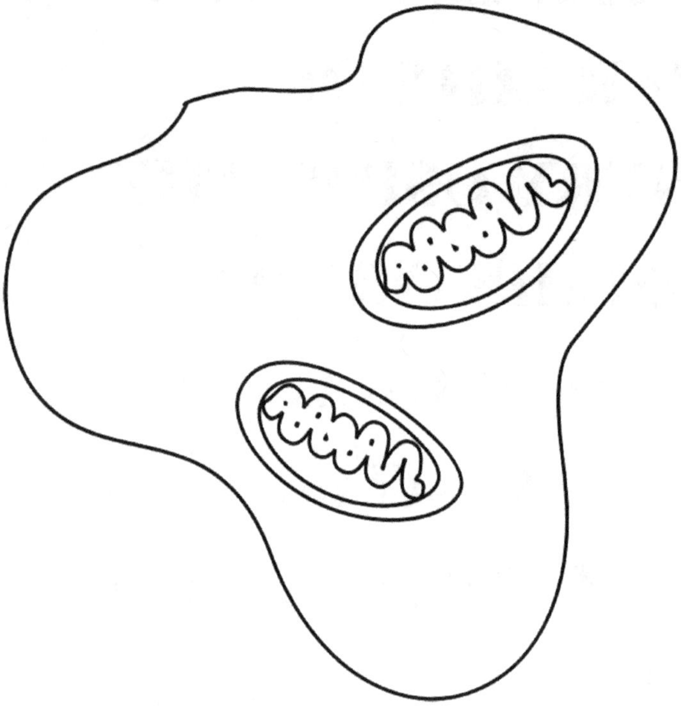

MITOCHONDRION IS A SMALL DOUBLE-MEMBRANE ORGANELLE FLOATING FREELY THROUGHOUT THE CELL.

IT FUNCTIONS AS THE POWERHOUSE OF THE CELL. IT GIVES THE CELL ENERGY IN A BIOCHEMICAL PROCESS THAT COMBINE FOOD (GLUCOSE) WITH OXYGEN TO PRODUCE ENERGY (ATP). THIS PROCESS IS CALLED CELLULAR RESPIRATION.

WE CAN THE MITOCHONDRIA THE BATTERIES PACK OF THE CELLS. THEY ARE FOUND IN ALMOST EVERY CELL IN YOUR BODY. EVEN ANIMALS AND PLANTS HAVE THESE STRUCTURES.

Most DNA is packages in chromosomes within the nucleus. The mitochondria also have a small amount of their own DNA. This genetic material is known as Mitochondrial DNA or mt- DNA . DNA or genes located in the nucleus get shuffled every generation. This is called recombination. Genetic recombination is the production of offspring with combinations of traits that differ from those in either parent. However, mtDNA does not get shuffled and recombination does not take place.

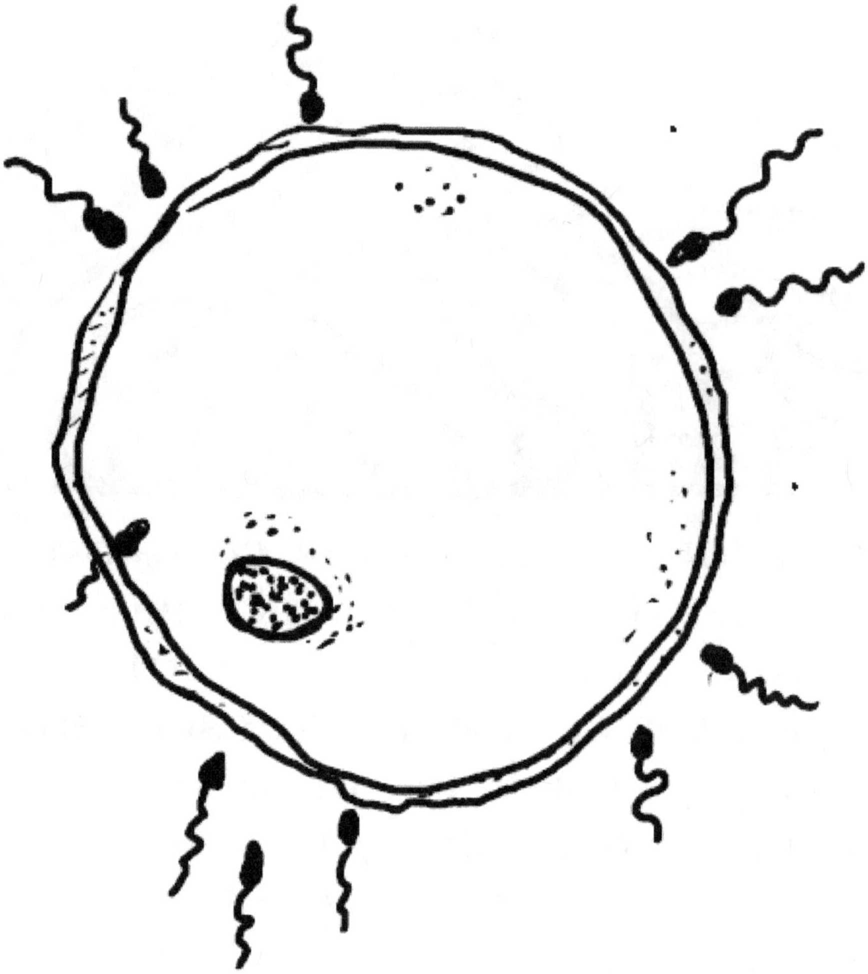

EVERY ORGANISM RECEIVES THEIR MITOCHONDRIA FROM THEIR MOTHER. THIS IS MADE POSSIBLE DURING THE TIME OF CONCEPTION. BEFORE CONCEPTION MITOCHONDRIA ARE IN THE EGG AND THE TAIL OF THE SPERM. THE SPERM NEED VAST AMOUNTS OF ENERGY TO REACH THE EGG FOR FERTILIZATION. UPON PENETRATING THE EGG CELL MEMBRANE, THE TAIL IS SEAL OFF AND SPERM MITOCHONDRIA DOES NOT BECOME A PART OF THE NEWLY DEVELOPING EMBRYO.

THUS, YOU HAVE ONLY THE MATERNAL MITOCHONDRIA. A BIT OF EVE IN ALL OF US.

# One BLOOD for all Nations

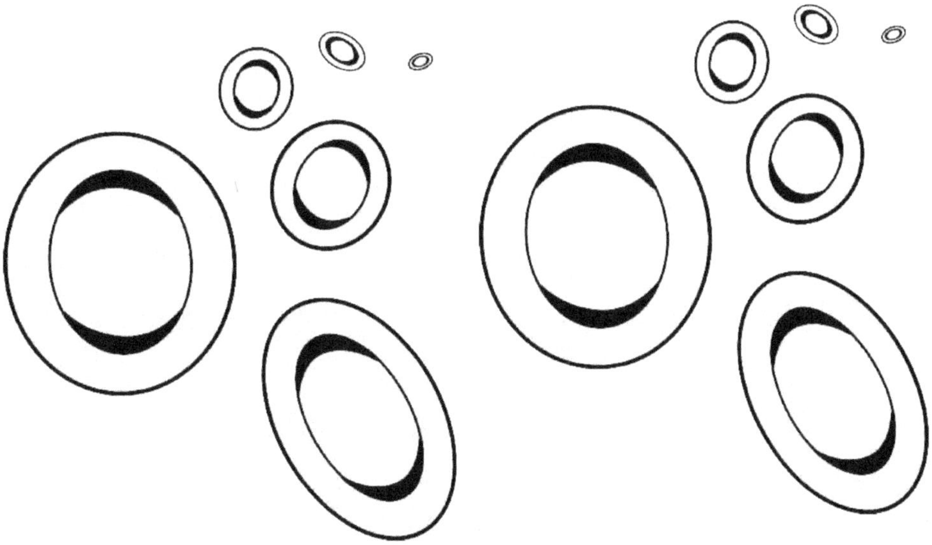

"The first two people of creation, Adam and Eve gave rise to all the ABO blood types in humans today. "And God made of one blood all nations of mankind to dwell on all the face of the Earth "

Acts 17: 26

# Hematopoiesis

## For the life of the flesh is in the blood- Leviticus 17:11

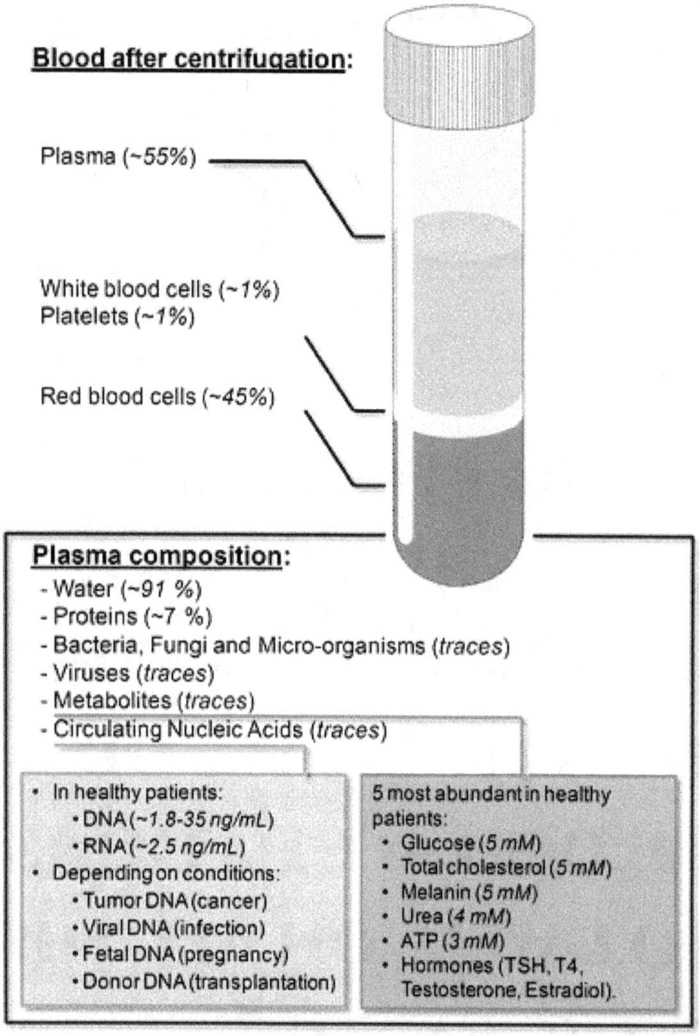


## Origin of Blood

**Blood is composed of formed elements. These elements are called red blood cells (erythrocytes), white blood cells (neutrophils, lymphocytes, eosinophils, basophils, and monocytes), and platelets (thrombocytes).**

- **Neutrophil** – engulf foreign debris
- **Red blood cell-** carrier oxygen
- **Platelets** – blood clotting
- **Eosinophil-** release histamine
- **Monocyte-** fight germs and clean up debris in would healing
- **Lymphocytes** – response to microorganisms and foreign bacteria, secretions of antibodies

# Formed Elements of Blood

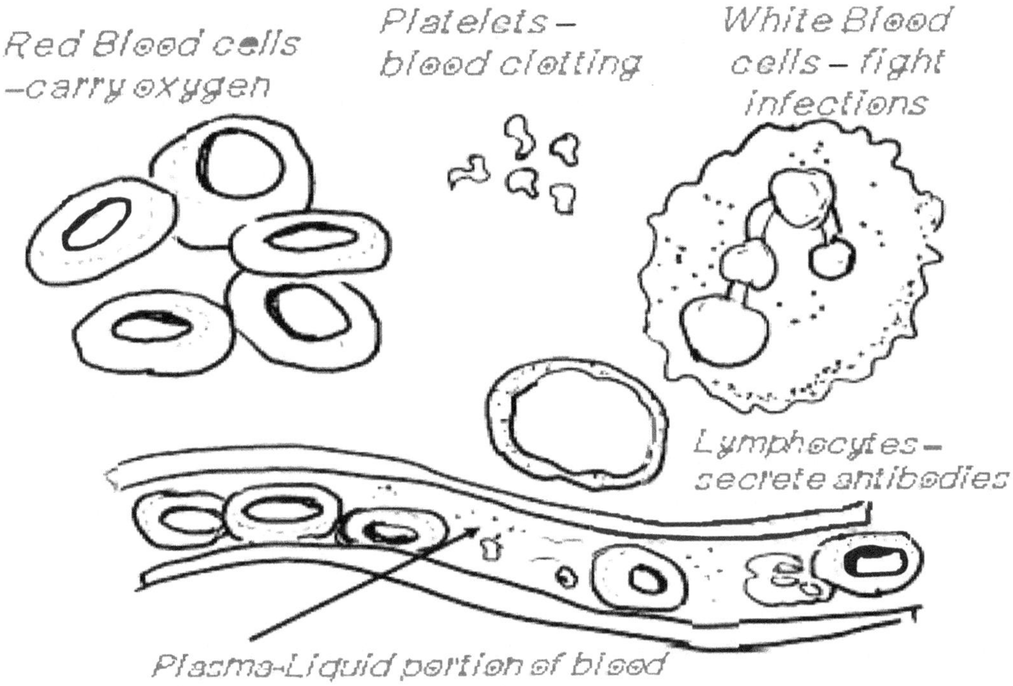

PLASMA IS THE LIQUID PORTION OF BLOOD. IT IS MADE UP OF 90% WATER WITH 10% IONS, SALTS, ENZYMES, DISSOLVED GASES, NUTRIENTS MOLECULES, AND WASTES. THE PROTEINS IN PLASMA INCLUDE ANTIBODY PROTEINS, CLOTTING FACTORS, AND ALBUMIN TO HELP KEEP THE RIGHT AMOUNT OF FLUIDS IN YOUR BLOODSTREAM

THE PROCESS BY WHICH BLOOD CELLS ARE FORMED IS CALLED HEMATOPOIESIS (*LATIN. HEMATO* -BLOOD, *POIESIS* - TO MAKE)

THE FORMATION OF BLOOD CELLS TAKES PLACE IN RED BONE MARROW FOUND IN THE END OF BONES. THESE EXTRACTIONS HAD THE POTENTIAL TO FORM BLOOD ELEMENTS IN USED IN THE CREATION OF EVE.

# Daughters of Zelophehad

**Daughters of a sonless father –**

**Numbers 27:1-11; 36:1-13**

# The Hemorrhaging Woman

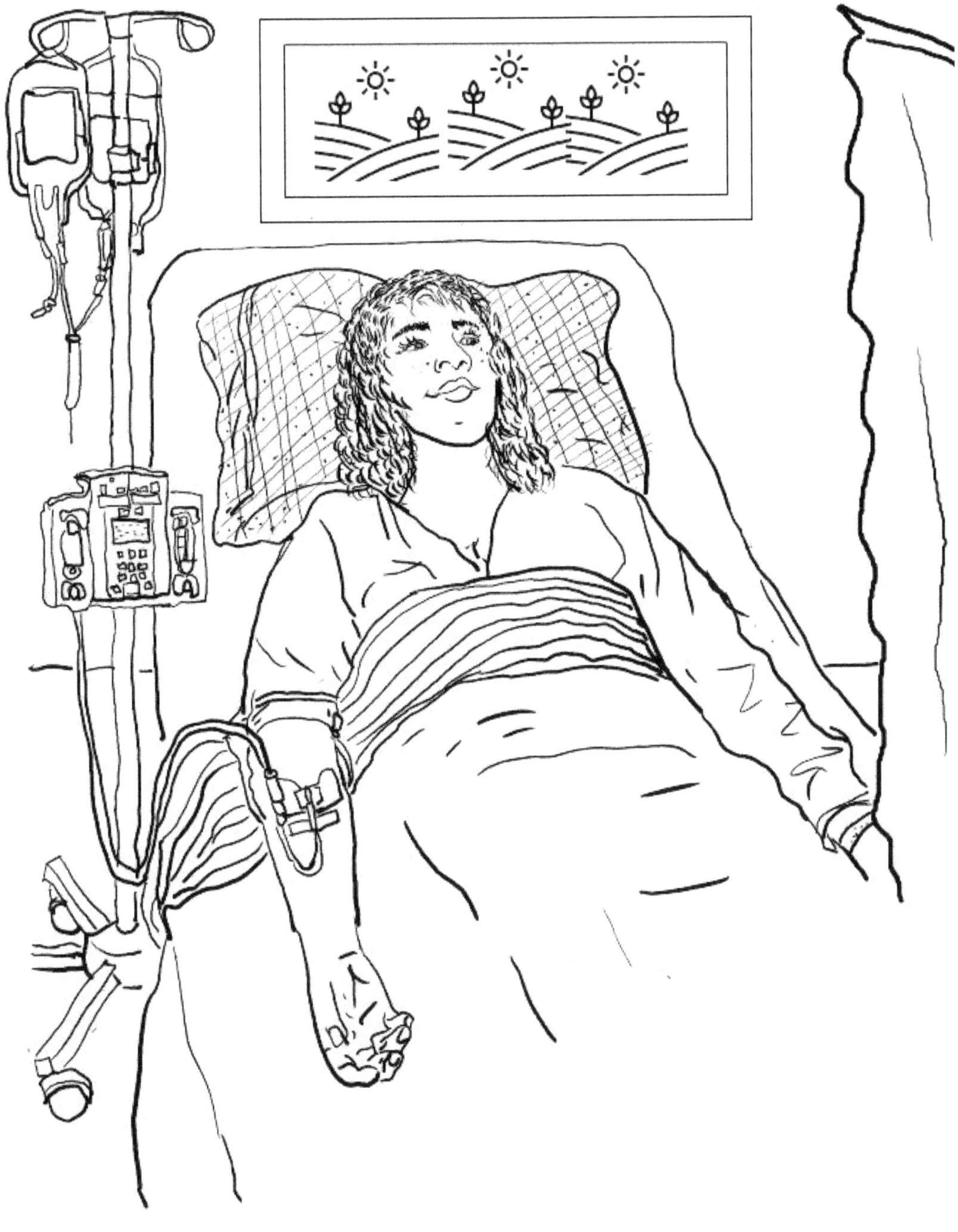

**She trusted in Jesus- power to heal**

**Matthew 9:20-22; Mark 5:25-34 ;**

**Luke 8:43-48**

# Journal Notes

# Functions of Blood

- Oxygen supplier to all cells
- Wound healer in clot formation
- Main transporter of nutrients – vitamins, sugar, electrolytes
- Maintain stable and constant body temperature
- Normalize pH levels in tissues
- Immune system defender – detection of foreign material by antibodies
- Hydraulic regulator – keep channel open for flow of ions
- Messenger – transport hormones, enzymes to target cells
- Removal of waste materials – carbon dioxide, urea, and lactic acids

# Blood Groups

A PERSON'S BLOOD TYPE IS DETERMINED BY THE PRESENCE OR ABSENCE OF SPECIFIC SURFACE STRUCTURES CALLED ANTIGENS. THEY ARE LOCATED ON THE RED BLOOD CELL SURFACE. ANTIGENS ARE SUBSTANCES MADE MAINLY OF PROTEINS THAT CAN TRIGGER AN IMMUNE RESPONSE TO DEFEND AND PROTECT THE BODY FROM INFECTIONS. ALL THE BODY CELLS IN AN INDIVIDUAL'S BODY HAS THE SAME PATTERN ON THE SURFACE OF THEIR RED BLOOD CELLS. THE DIFFERENT BLOOD GROUPS ARE A, B, AB, AND O.

| | Type A | Type B | Type AB | Type O |
|---|---|---|---|---|
| **Antigen** (on RBC) | Antigen A | Antigen B | Antigens A + B | Neither A or B |
| **Antibody** (in plasma) | Anti-B Antibody | Anti-A Antibody | Neither Antibody | Both Antibodies |
| **Blood Donors** | Cannot have B or AB blood<br>Can have A or O blood | Cannot have A or AB blood<br>Can have B or O blood | Can have any type of blood<br>Is the universal recipient | Can only have O blood<br>Is the universal donor |

# Genetics-

**IS THE STUDY OF GENES THAT ARE THE BASIC UNIT OF HEREDITY.**

**A PUNNETT SQUARE IS A GRID OR DIAGRAM USED TO PREDICT THE PROBABILITY OF AN OFFSPRING'S GENOTYPE (THE GENE COMBINATION OF AN ORGANISM RECEIVED FROM THEIR MOM AND DAD).**

**SCIENTIST STUDY GENETICS TO BETTER UNDERSTAND HOW TRAITS ARE INHERITED. THE POSSIBLE INHERITANCE OF HUMANKIND FOUR BLOOD TYPES FROM ADAM AND EVE ARE A, AB, B, AND O.**

| | | Adam | |
|---|---|---|---|
| | | $I^A$ | $i$ |
| EVE | $I^B$ | $I^A I^A$<br>Seth?-"AB" | $I^B i$<br>Cain? "B" |
| | $i$ | $I^A i$<br>Abel?- "A" | $i\,i$<br>other sons and daughter- "O" |

# Procreative Principles-

Everything created by God was given the power to reproduce its 'own kind'.

Adam and Eve were the source from which humankind would continue.

The term 'Its own kind" is when individuals or groups share common traits in their chromosomes.

The power of reproduction is not in the egg or sperm nor the embryo. That power is only in the mature parents. In the beginning, the mature parents of all living things were created with eggs, seeds, bud, spores, and sperms within them for future multiplication on the Earth.

Living things cannot naturally break this law of reproduction and produce any other kind of organisms.

# Jezebel – 1 Kings 16:31; 18:1-13; 21:125

# Journal Notes

# The Seed of a Woman -Ovum

Genesis 3: 15 ; Genesis 4:1

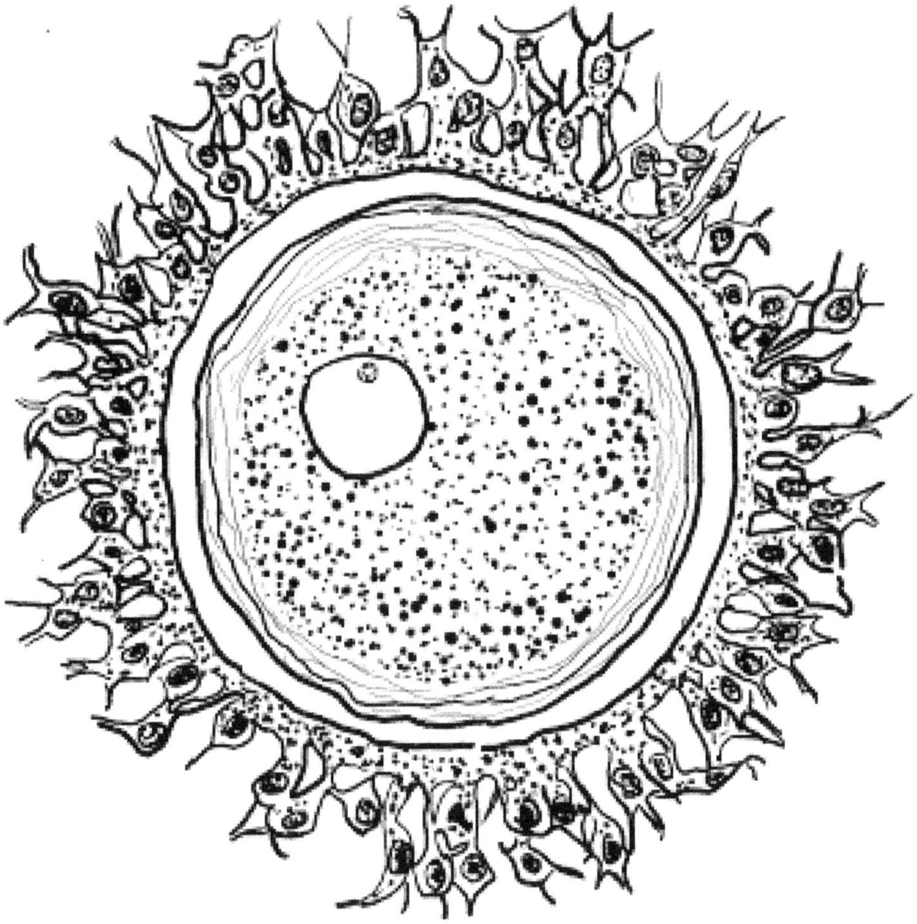

**THE EGG OR OVUM IS THE FEMALE REPRODUCTIVE CELL OR GAMETE. A GAMETE IS A HAPLOID SEX CELL THAT CONTAIN 23 CHROMOSOMES IN HUMANS. THE HUMAN EGG IS THE BIGGEST CELL IN THE BODY AND LARGER THAN THE MALE SPERM. IT IS ABOUT THE SIZE OF A SESAME SEED. ITS SIZE HELPS TO STORE THE NUTRIENTS FOR NEW LIFE.**

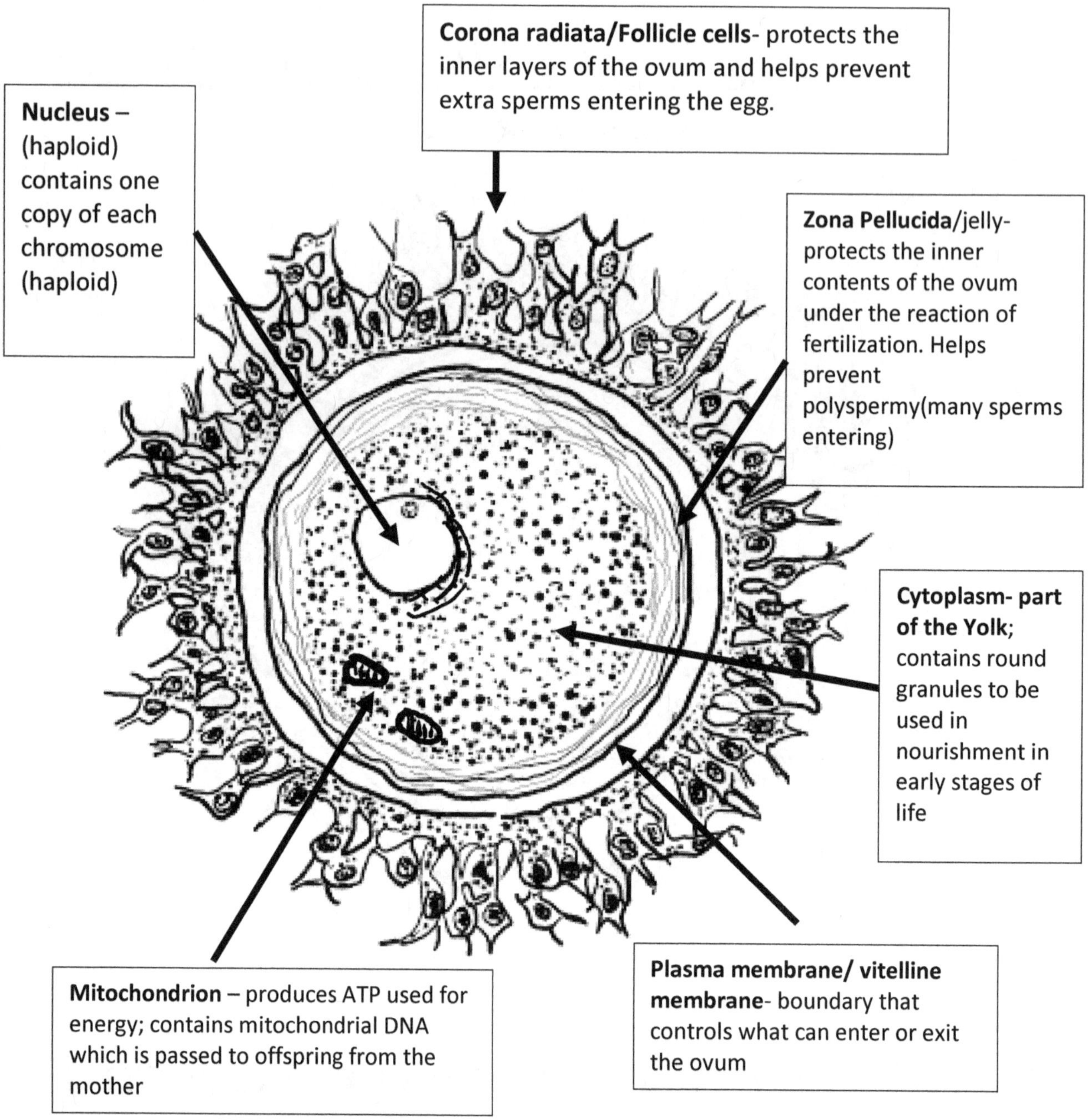


**THE FUNCTIONS OF THE OVUM ARE TO CARRY THE SET OF 23 CHROMOSOMES, TO PROVIDE THE NUTRIENTS FOR GROWING THE EMBRYO UNTIL IT SINKS INTO THE UTERUS WALL, AND TO DEVELOP THE PLACENTA AND TAKE THE NUTRIENTS FROM THE MOTHER'S BLOOD STREAM.**

EGGS ARE PRODUCED IN THE OVARY AND TRAVELS DOWN THE FALLOPIAN TUBES. THE EGGS REMAIN INSIDE A FEMALE OVARY FROM BIRTH. WHEN AN EGG IS RELEASED ONCE A MONTH IT LIVES ABOUT 12-24 HOURS AND THEN DIES IF NOT FERTILIZED. THE RELEASE OF AN EGG IS CALLED THE OVULATION CYCLE.

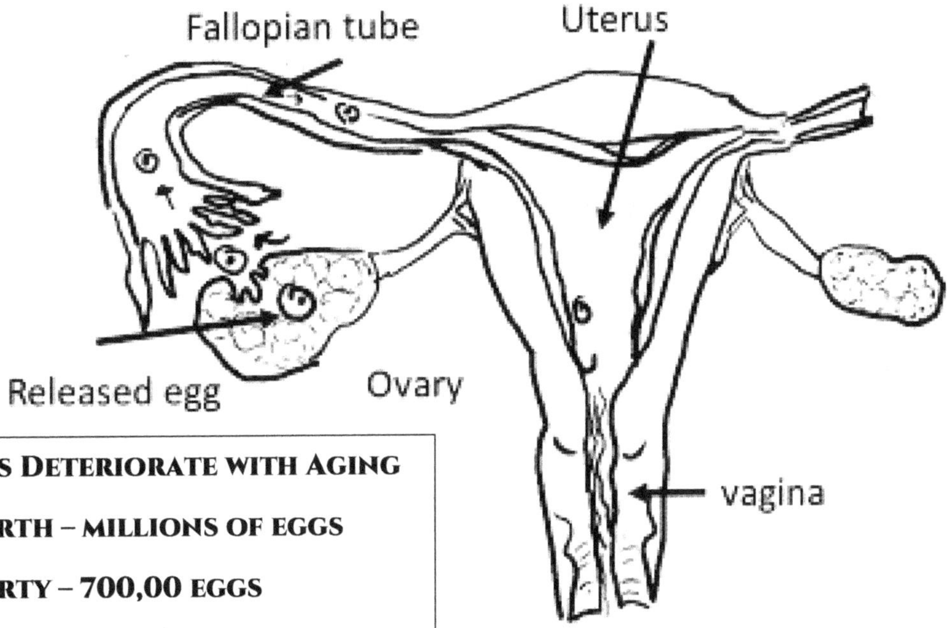


| EGGS DETERIORATE WITH AGING |
| --- |
| AT BIRTH – MILLIONS OF EGGS |
| PUBERTY – 700,00 EGGS |
| MENOPAUSE – 1,000 EGGS |

WHEN THE EGG AND SPERM FUSE DURING FERTILIZATION, A DIPLOID CELL (46 CHROMOSOMES) IS FORMED AND GROWS INTO A NEW ORGANISM. AT THE BEGINNING OF CREATION EVE'S OVARIES CONTAINED ALL THE OVUM TO HELP HER FULFILL THE COMMANDMENT TO, 'BE FRUITFUL, MULTIPLY, AND REPLENISH THE EARTH'- GENESIS 1:28.

# Anatomically designed to be a Life Bearer

**Adam saw his nature in Eve and called her womb-man (woman). Because she was taken out of Man -Genesis 2:23b**

# The Womb-Man = Woman

**Eve, the womb-man, carried many children. God created the morphology of Eve's body to prepare for life within her womb.**

WOMEN ARE BORN WITH ALL THEIR EGGS FOR A FERTILE LIFETIME. THIS MEANS YOUR EGGS WERE ONCE INSIDE YOUR MOTHER WHEN YOU WERE A FETUS IN HER WOMB. THAT MEANS YOUR MOTHER CARRIED THE EGG CELLS WHICH MAY ONE DAY BE FERTILIZED AND GROW INTO HER OWN GRANDCHILD. GENESIS 5:4

*www.naturalcycles.com*

# Second Generation Christians- Acts 2:17

**Philip's Daughters**

# Neuroscience

## The structure and functions of the nervous system and the brain.

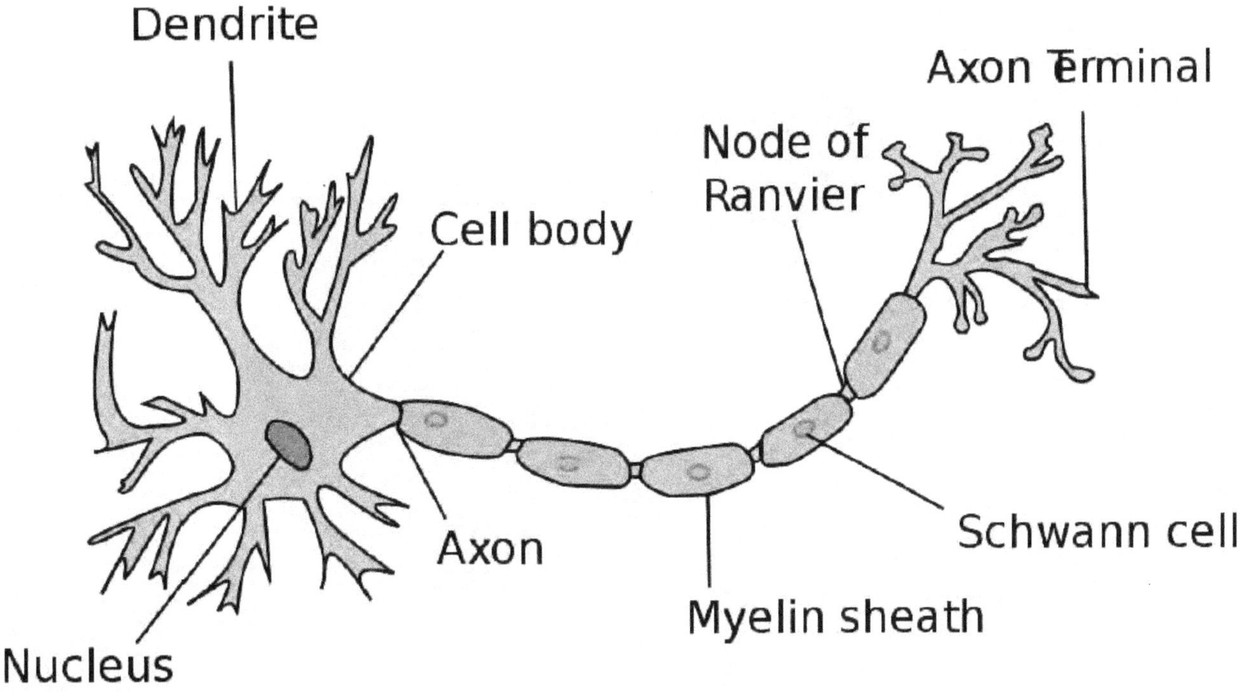


**Neurons are nerve cells that make up the brain. They are electrically excitable cells. Neurons send and receive signals from your brain up to 200 mph. In humans there are an estimated 10-20 billion neurons in the front area of your brain.**

# Eve, the Thinking Woman

## FUNCTIONAL DIFFERENCES IN THE FEMALE BRAIN

When listening neurons on both sides of the brain are activated

Higher sense Of smell

Better memory for details, both pleasant and unpleasant

more taste buds on tongue

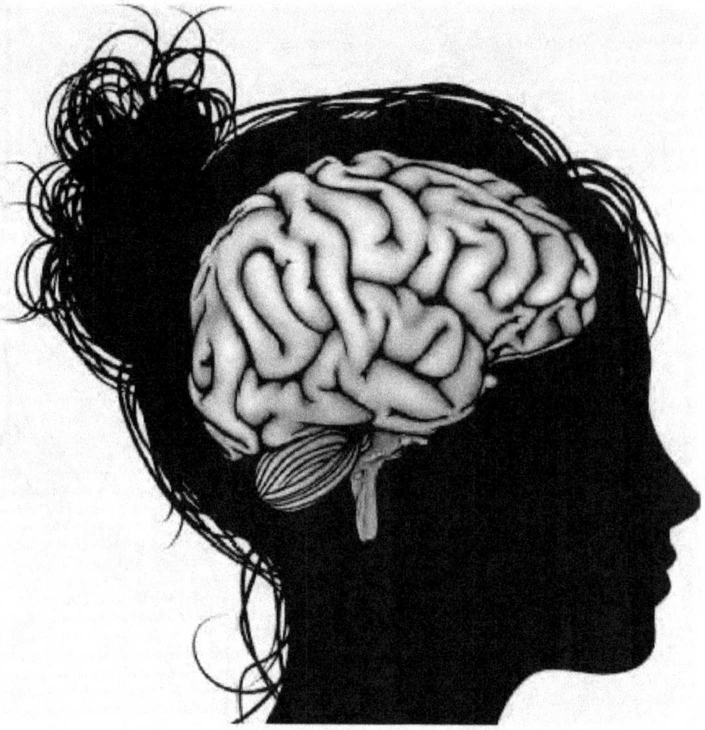

High language process center

More detailed memory

Larger pupils that allows for more light and help see more colors

More able to express emotions

Can recall facial recognitio n faster

More white brain matter for processing information

Has a more empathizing brain, wired to nurture, and connect

Good at problem solving and decision-making

**THE BRAIN IS EXTRAORDINARILY COMPLEX. THE MINDS OF MEN AND WOMEN ARE 99% THE SAME, BUT THE 1 % MAKES THE DIFFERENCE!**

THE BRAIN IS ABOUT THREE POUNDS AND IS ONE OF THE MOST POWERFUL ORGANS CONTROLLING ALL THE BODY. IT IS LIKE A LIVING COMPUTER COMPOSED OF 100 BILLION NERVE CELLS. THE BRAIN ASSEMBLES AND INTERPRETS INFORMATION RECEIVED FROM THE SENSE ORGANS AND THE MUSCLES.

IT IS ALSO THE CENTER OF EMOTIONS AND INTELLECT. THE BRAIN ASSEMBLES AND INTERPRETS INFORMATION RECEIVED FROM THE SENSE ORGANS AND THE MUSCLES.

THE THREE MAIN PARTS OF THE BRAIN ARE CEREBRUM, CEREBELLUM, AND THE BRAIN STEM.

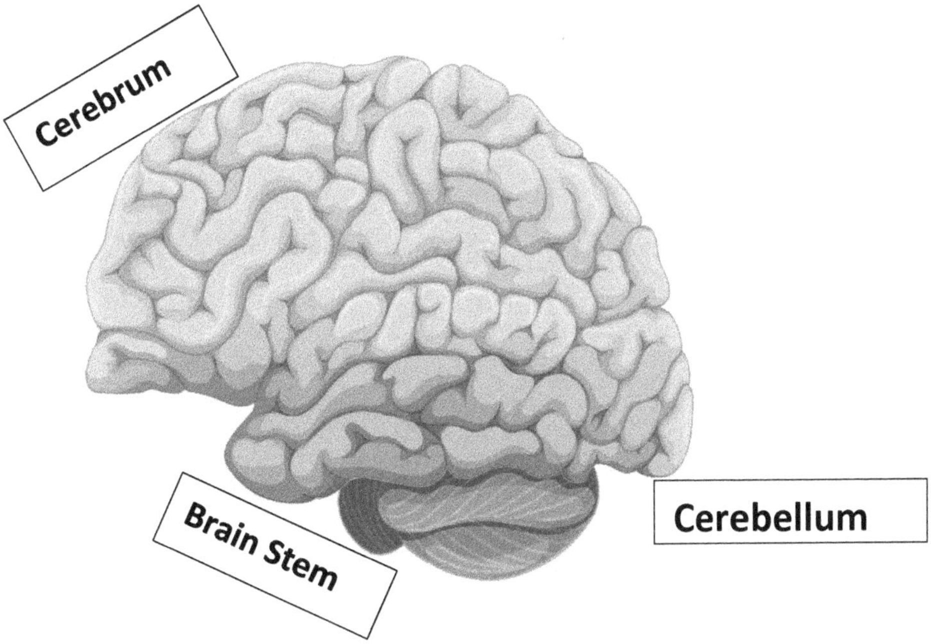


- CEREBRUM IS INVOLVED IN REMEMBERING, PROBLEM, SOLVING, THINKING, FEELING AND CONTROLS MOVEMENT
- CEREBELLUM CONTROLS COORDINATION AND BALANCE
- BRAIN STEM CONNECTS THE BRAIN TO THE SPINAL CORD AND CONTROLS AUTOMATIC FUNCTIONS SUCH AS BREATHING, DIGESTION, HEART RATE, AND BLOOD PRESSURE.

# Cerebral Hemispheres

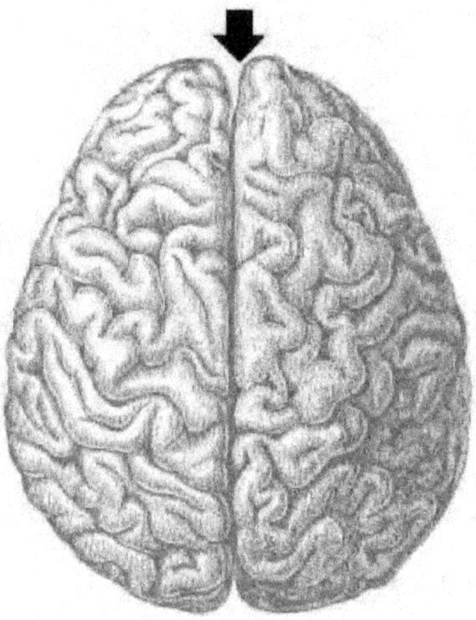


| | |
|---|---|
| Language | Control the muscles on the left side of the body |
| Logical thinking | Comprehend visual imagery |
| Mathematical operation | Understanding abstract concepts |
| Detailed skeletal motor control | Creativity / Imagination |
| Processing of fine visual and auditory details | Memory, Non-verbal thinking |
| Controls the muscles on the right side of the body | Spatial relationship and patterns |
| Processing what you hear | Face recognition |

**The brain is divided into two sides or hemispheres. Each side can be divided into four lobes: frontal, temporal, parietal, and occipital. Men generally are prone to be more left-brain strong. They approach problem-solving from a task-oriented perspective and women solved problems through feelings, communication, and creativity.**

**HERODIAS & HER DAUGHTER -**
**MATTHEW 14:1-12; MARK 6:14-29**

# How the Brain Works

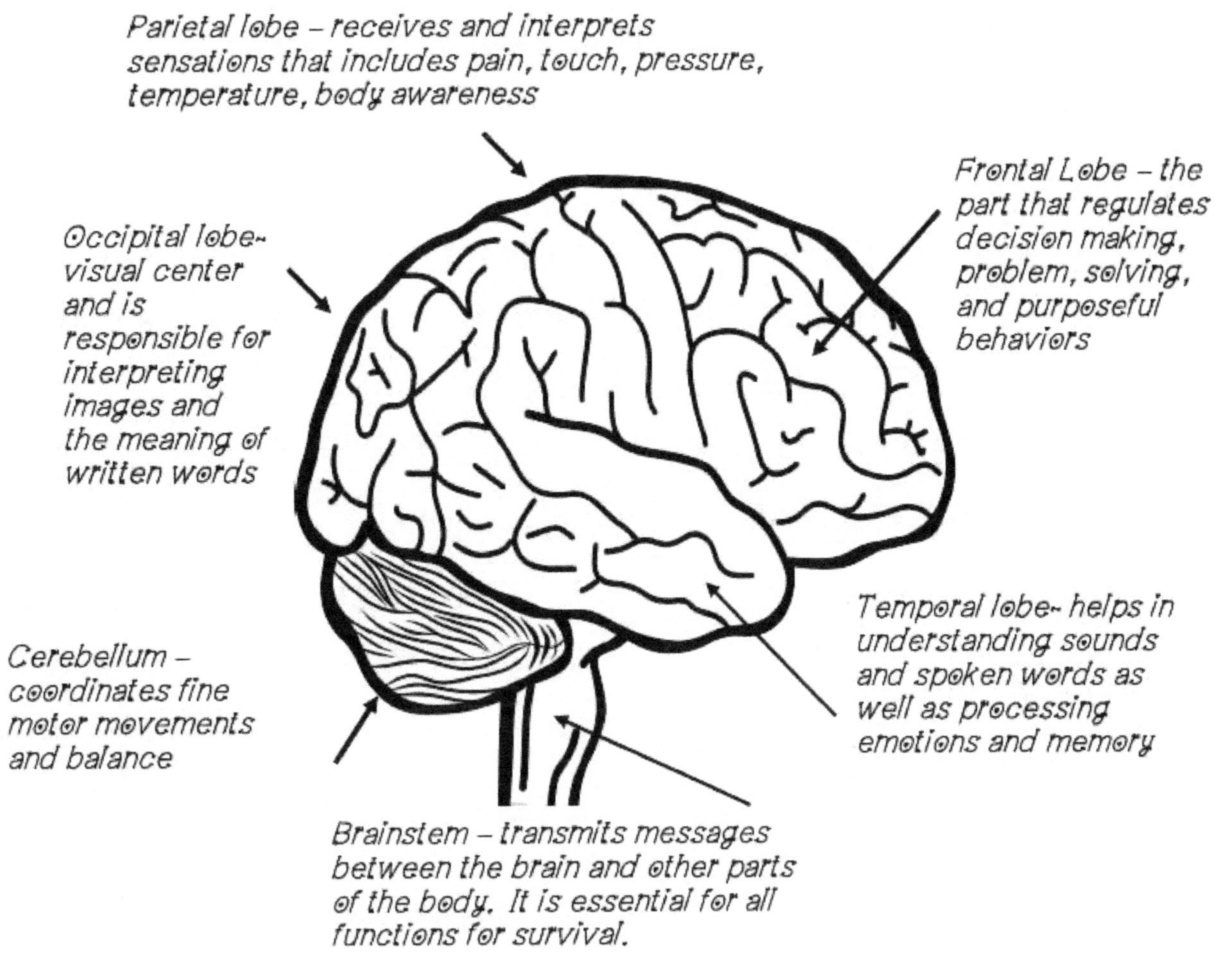


**Scientists have discovered that there are actual structural differences between Adam (male) and Eve (female) brains. Certain parts or areas of their brains are slightly different. Women tend to communicate more effectively than men They talk through issues, express emotions, and are more sociable in groups. Men are more task oriented. Adam was given the assignment of tending the Garden of Eden and naming the animals (Genesis 2:8). Men are less talkative. They tend to process best in the left hemisphere while women tend to process equally between right and left hemispheres.**

# Endocrine System

## Hormone Messengers

**HORMONES ARE CHEMICAL MESSENGERS THAT INITIATES AND CONTROLS VITAL BODY RESPONSES. THE ENDOCRINE SYSTEM REGULATED BY THE PITUITARY GLAND LOCATED AT THE BASE OF THE BRAIN. THIS SYSTEM IS RESPONSIBLE FOR REGULATING GROWTH, METABOLISM, REPRODUCTIVE PROCESSES, HOMEOSTASIS, TISSUE DEVELOPMENT, SEXUAL FUNCTIONS, AND BEHAVIOR.**

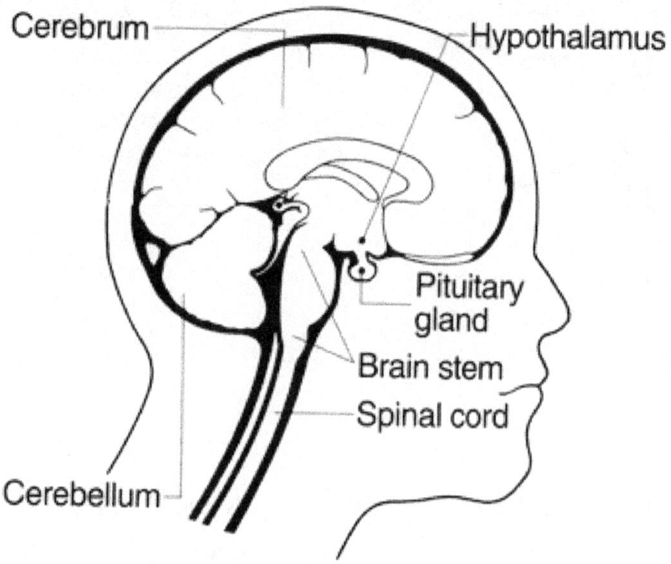


**THE PITUITARY GLAND RELEASES HORMONES TO THE MALE AND FEMALE REPRODUCTIVE ORGANS. THE TWO MAIN SEX HORMONES ARE TESTOSTERONE PREDOMINATELY FOUND IN MALES AND ESTROGEN AND PROGESTERONE IS PREDOMINANTLY FOUND IN FEMALES.**

**TESTOSTERONE ALLOWED FOR MALES (ADAM) GREATER MUSCLE MASS, LARGER BONES, STRONGER LIGAMENT, AND TENDONS THAN POSSIBLY FOUND IN FEMALES (EVE). THE MALE'S BODY CAN ALSO PRODUCE MORE RED BLOOD CELLS WHICH ARE NEEDED TO DELIVER MORE OXYGEN TO MASSIVE MUSCLES FOR STRENUOUS WORK (GENESIS 2:8,15).**

# Important Hormone for Eve's Good Health

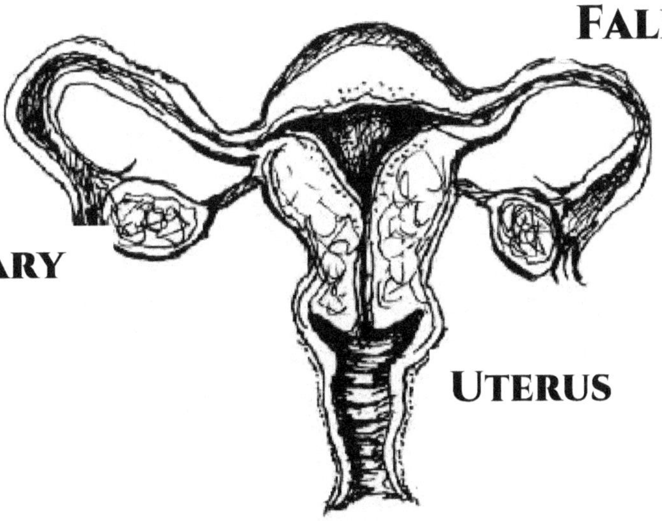


ESTROGEN IS PRODUCED AND RELEASED PRIMARILY BY THE OVARIES. ESTROGEN AND PROGESTERONE PREPARED EVE'S BODY FOR PREGNANCY (GENESIS 4:1-2; 4:25-26) , ALLOWED FOR SECRETION OF OSTEOBLASTS (BONE FORMING CELLS) THAT BINDS WITH CALCIUM AND VITAMIN D, HELPS THE HEART BY RELAXING BLOOD VESSELS TO INCREASE BLOOD FLOW, HELPS TO KEEP THE GOOD CHOLESTEROL LEVEL HIGH REGULATES FLUID BALANCE, INCREASE FAT STORAGE, AND HELPS IN SYNTHESIS OF PROTEINS.

PROGESTERONE LEVELS RISE AFTER OVULATION AND PREPARE UTERUS FOR A FERTILIZED EGG. IF PREGNANCY DOES NOT HAPPEN THE PROGESTERONE LEVEL WILL DROP, AND YOUR MENSTRUAL CYCLE WILL BEGIN.

PROLACTIN IS RELEASED AFTER CHILDBIRTH TO TRIGGER LACTATION WHICH ENABLES WOMEN TO BREASTFEED.

TESTOSTERONE, A LITTLE OF THIS MALE HORMONE IS IN FEMALES AS WELL. IT PLAYS A ROLE IN YOUR SEX DRIVE, BONE DENSITY, AND MUSCLE STRENGTH.

# Journal Notes

Hagar - She wept and God saved her son- Genesis 16:13

THE FIRST PERSON IN THE BIBLE TO CRY

# Journal Notes

# Adipose Tissues

## 'the curvy woman'

Adipose tissues are also called fat tissues. Their main role is to store energy in the form of lipids. These cells swell as they store fat and shrink when the fat is used for energy.

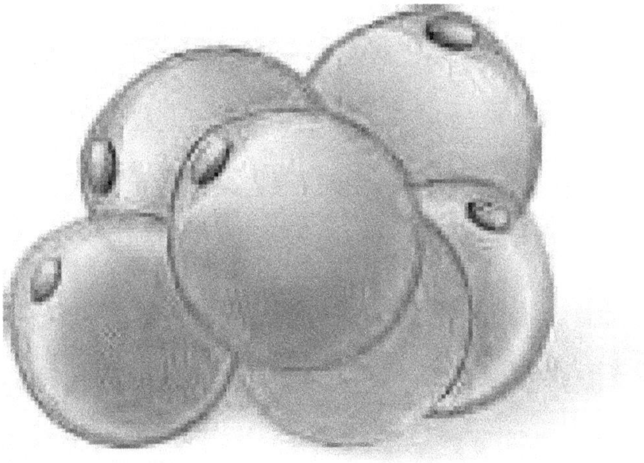

Many negative thoughts are associated with the distribution and deposit of adipose tissues on a woman's body. It is important to note that people cannot thrive without fat cells.

# Functions of Adipose Tissues

- Fats cells also produce hormone like estrogen which help in the regulation of vital bodily functions.
- Major ingredient in brain tissue
- A structural part of all cell membranes
- Supplies a padding to protect internal organs against trauma to their tissues
- Soften the outline of muscles
- Protect nerve tissues and help in transmission of electrical impulses
- Helps regulate a woman's menstrual cycle
- Insulates the body against the cold
- Rounds out the body's surfaces while creating characteristics folds and grooves

As most woman age there is a decline in female hormone levels. This can accelerate body fat accumulation. This is a good reason to keep exercising for your better health.

Eve, compared to Adam, probably had a higher percentage of adipose tissue or body fat (18-20%). She most likely had a round and curvy body.

# THE CURVY WOMAN

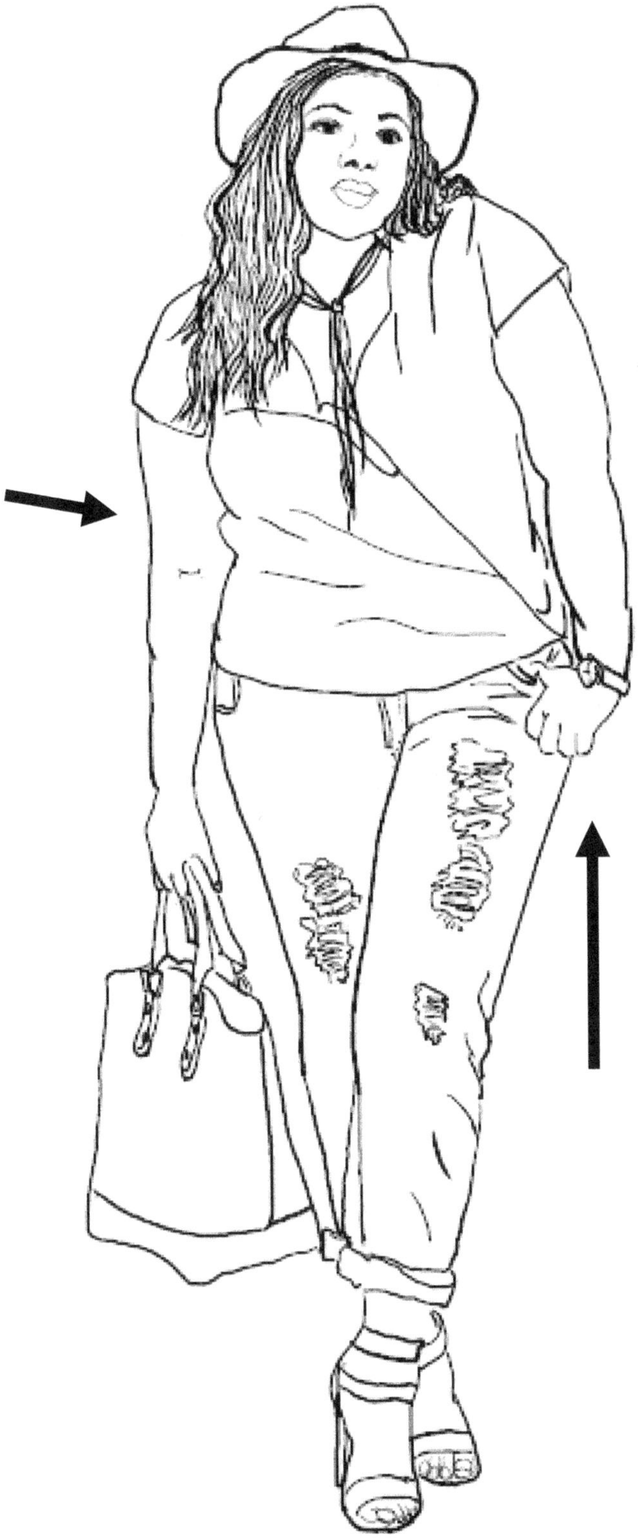

DISTRIBUTION OF FAT CELLS CAN BE FOUND IN SEVERAL PLACES ON THE FEMALE BODY. FATS CELLS ARE LOCATED ON THE BUTTOCKS; BETWEEN THIGHS, HIPS, PUBLIC AREA, KNEES; THE BREAST , PARTS OF THE UPPER ARMS AND AROUND HER NAVEL ADIPOSE TISSUE CAN BE FOUND BETWEEN MUSCLES; AROUND INTERNAL ORGANS TO CUSHION AND PROTECT AND ALSO INSULATES THE BODY FROM HEAT LOSS . THE FEMALE BODY STOCKS ENERGY IN THE FORM OF FAT IN ANTICIPATION OF FUTURE PREGNANCIES.

ADIPOSE TISSUE AID IN THE SURVIVAL OF THE SPECIES AS IT PROVIDES FOR SURVIVAL OF THE WOMAN AND HER OFFSPRING DURING TIME OF SCARCITY OR FAMINE.

# Journal Notes

# Bone of My Bones

# Journal Notes

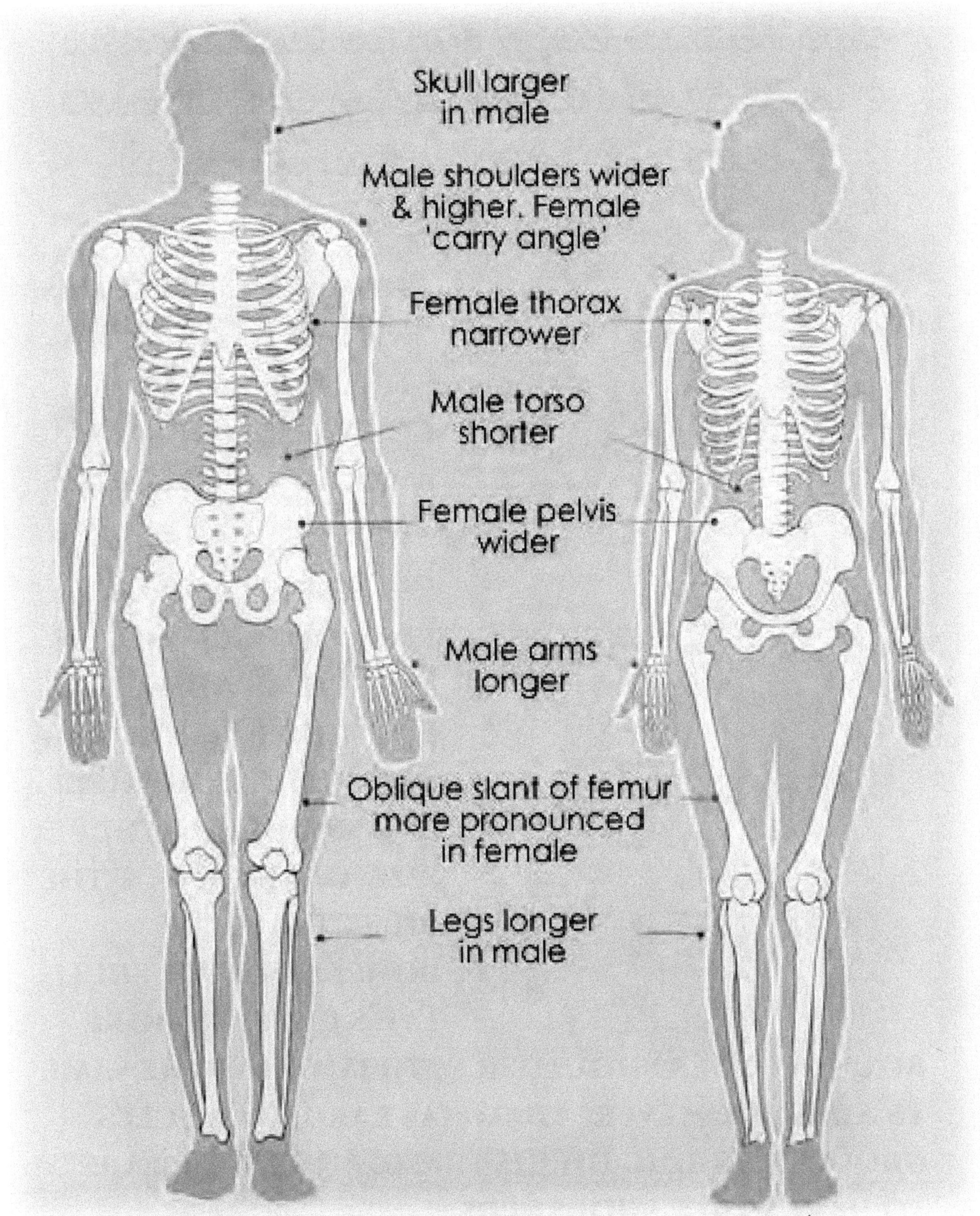
Skull larger
in male
Male shoulders wider
& higher. Female
'carry angle'
Female thorax
narrower
Male torso
shorter
Female pelvis
wider
Male arms
longer
Oblique slant of femur
more pronounced
in female
Legs longer
in male

**Genesis 2:23**

## The Female Body Shape is the total quantity of distribution of muscles tissues, fat cells, and skeletal structure

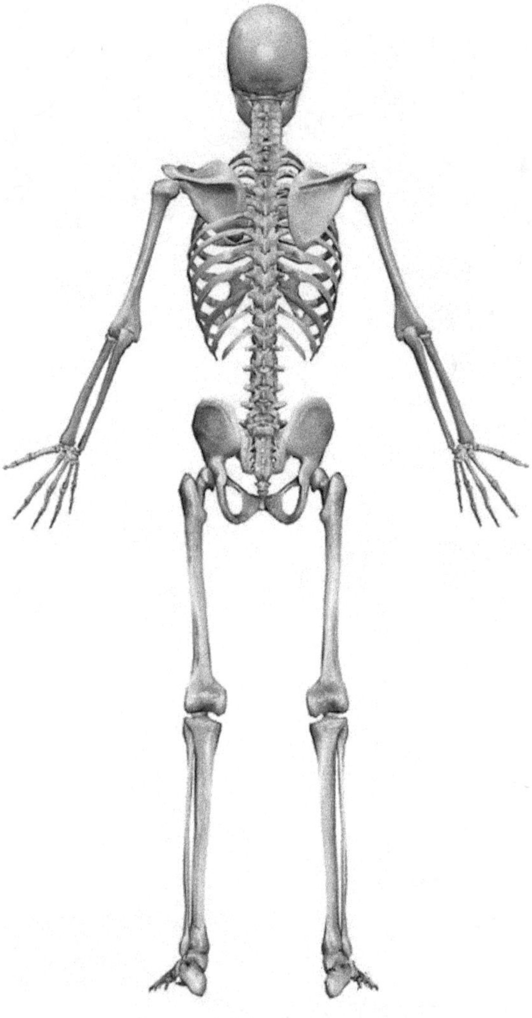

THE MALE SKELETON IS THICKER AND STRONGER AND CONTAINS A HIGHER AMOUNT OF CALCIUM THAN FEMALES.

EVE'S SKELETON OR FRAMEWORK IS NOT AS MASSIVE AS ADAM'S. THE THREE MAJOR DIFFERENCES IN THEIR SKELETONS ARE THE SIZE AND SHAPE OF THE RIB CAGE, PELVIS BONES, AND THE SKULL. EVE'S CHEST IS MORE ROUNDED. HER SHOULDER WITH IS ABOUT THE SAME AS ADAM. HOWEVER, ADAM HAS LARGER MUSCLES GROUPS HANGING ON HIS FRAME AND HE APPEARS TO HAVE WIDER SHOULDERS.

**THE LUMBAR CURVE IS GREATER IN WOMEN.**

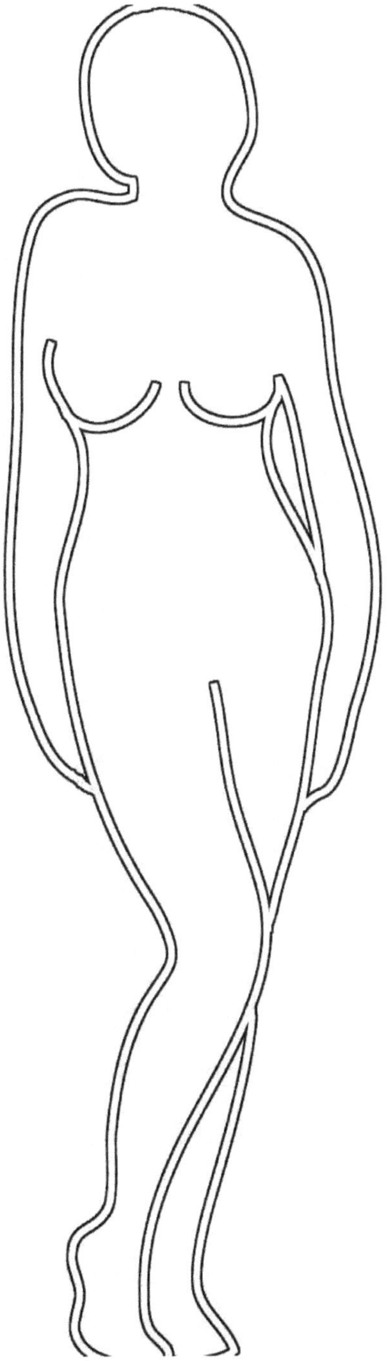

**HER PELVIS IS TILTED FORWARDS WHICH MAKES FOR THE S-SHAPED APPEARANCE IN HER BACK. THE FEMALE FIGURE IS TYPICALLY NARROWER AT THE WAIST THAN AT THE BUST AND HIPS. EVE'S PELVIS WOULD BE WIDER AND MORE CIRCULAR TO ADAPT FOR GESTATION AND FOR THE PASSAGE OF A BABY DURING BIRTH. HER HIPS ARE WIDER BECAUSE OF THE POSITION OF THE FEMUR.**

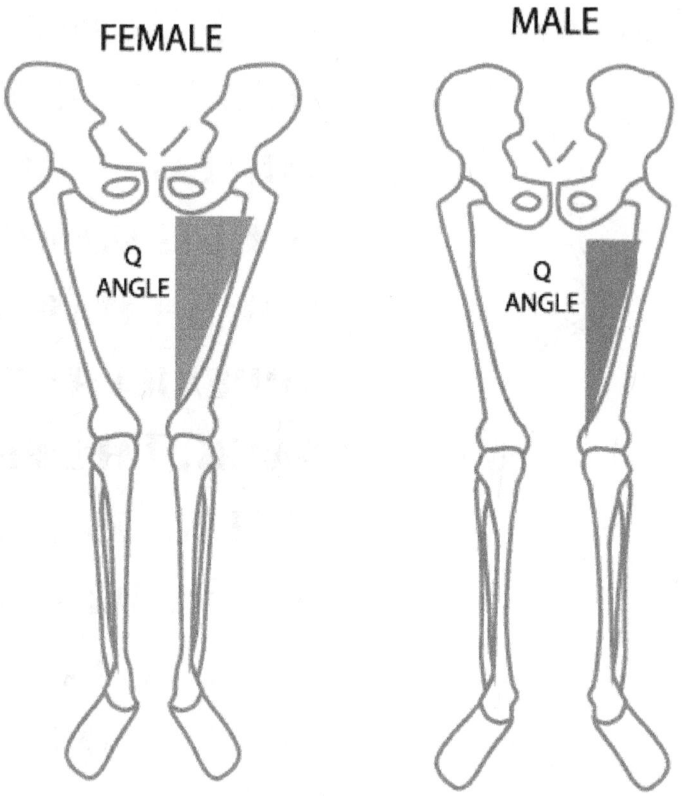


**Eve's femur is more angled than in Adam's long leg bone. This structural arrangement gives Eve a slightly "X" shaped frame.**

**Women have looser ligaments (a band of tough, flexible connective tissue) which connects bone to bone. A women's slanted thigh bones put extra pressure on the knee joints, which must rotate while men's do not.**

**A woman would have a greater endurance or steady power to run a race while a man would have a great explosive power but fatigues more easily. Her ligaments provide for greater flexibility in joints.**

# Bilateral Symmetry and how it help you

Symmetry, form, size, arrangement, shape, balance, and proportion of parts on opposite sides of a dividing line. Psalm 139:13

# Benefits of Bilateral Symmetry

## Bilateral Symmetry helps You:

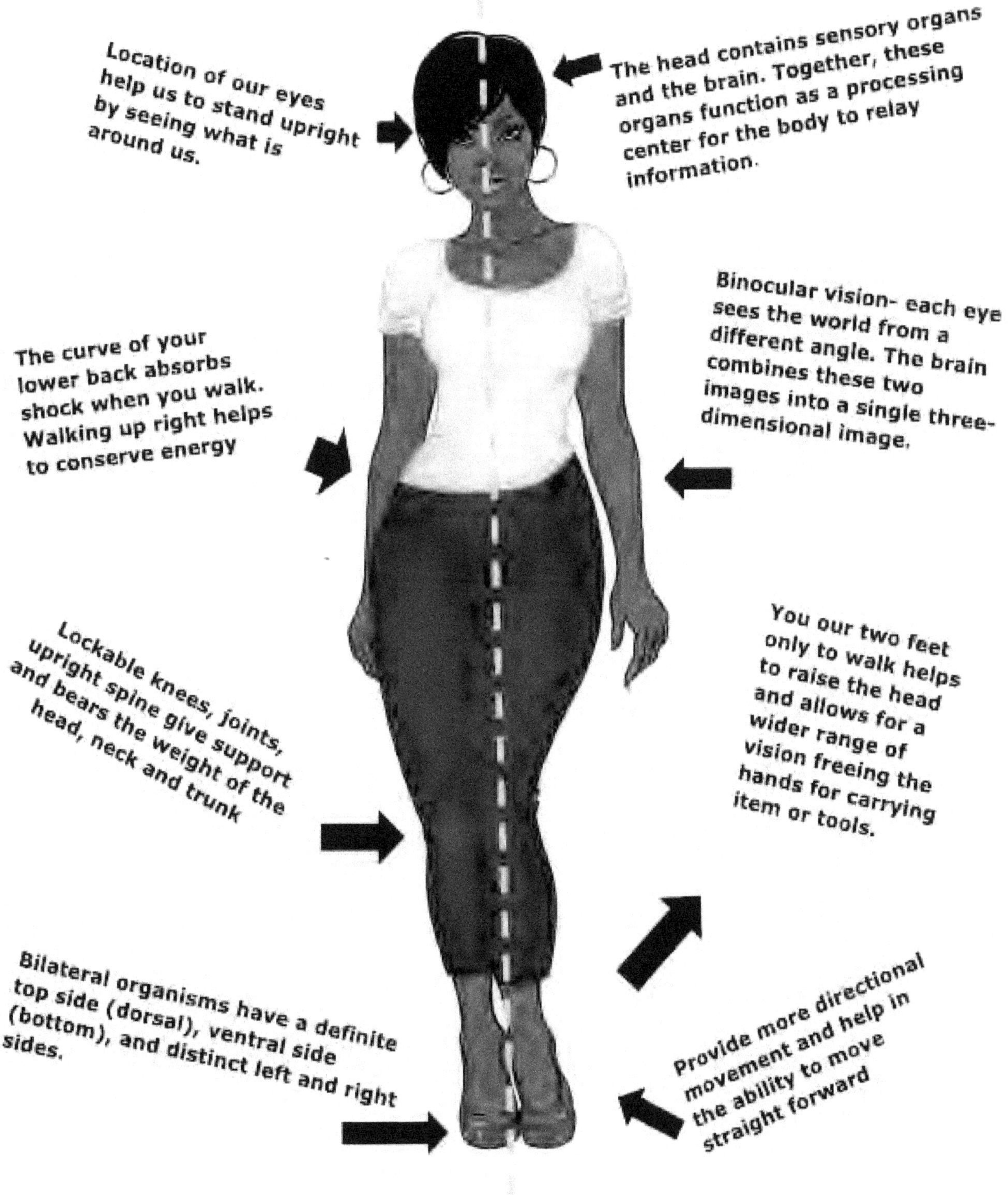

# Biomass

**WHAT DID EVE LOOK LIKE? WAS SHE CURVY, STRAIGHT, ROBUST, THIN, TALL? AS FOR HER GENERAL APPEARANCE, SHE WAS CREATED IN THE IMAGE OF GOD AND DESIGNED FOR PURPOSE AND SERVICE.**

# Sense Organs

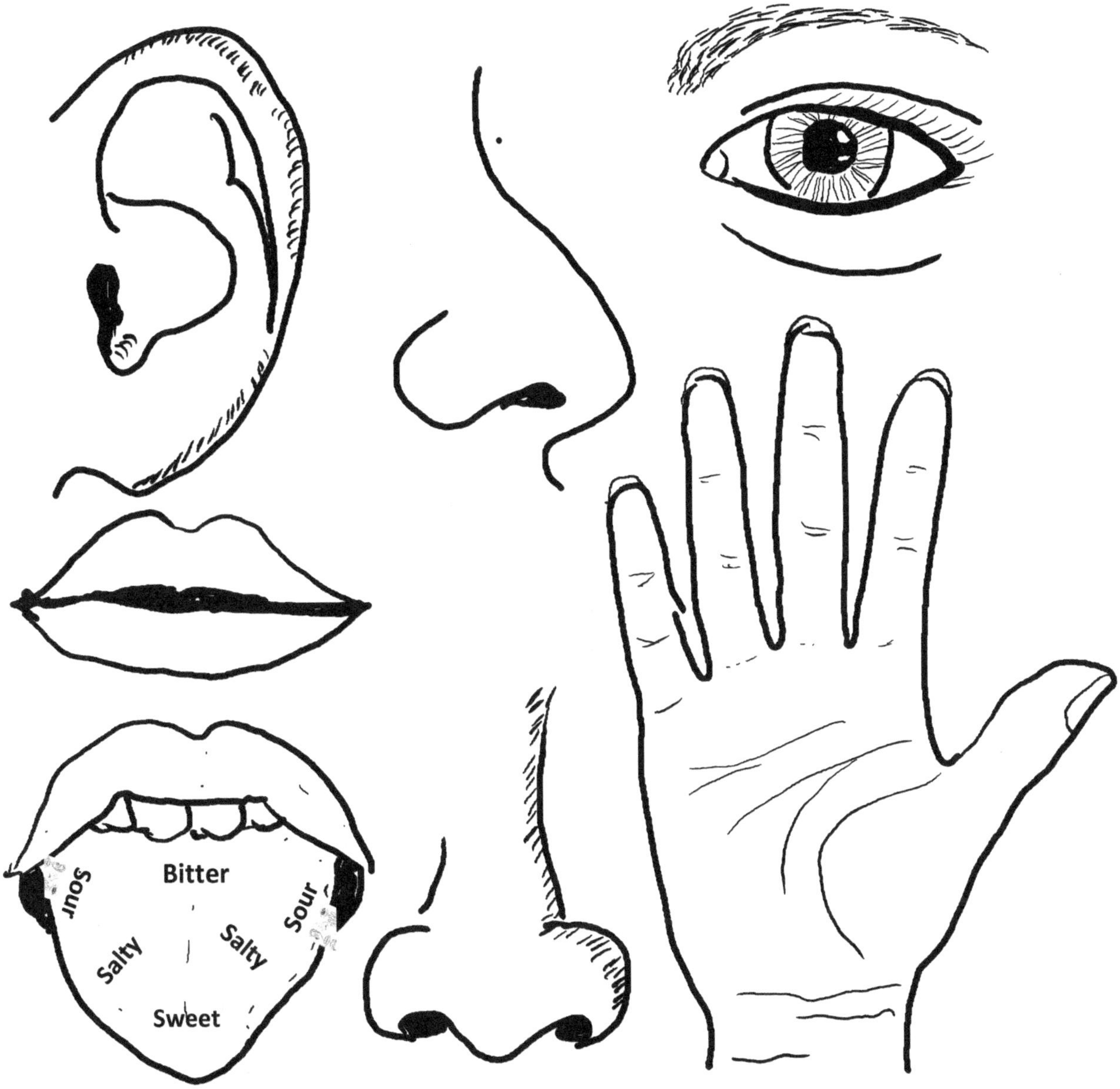


ADAM AND EVE USED THEIR SENSORY ORGANS TO SELECT AND EAT FOOD. WITHOUT THE SENSE OF TASTE MANY THINGS IN LIFE WOULD NOT BE AS PLEASANT.

# You Eat with Your EYES

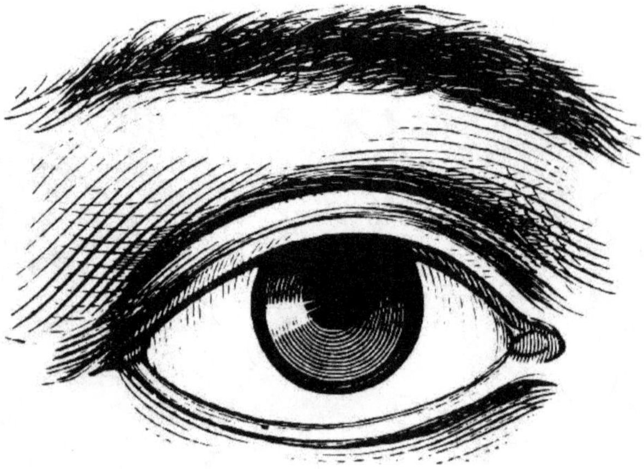

**AND WHEN THE WOMAN SAW THAT THE TREE WAS GOOD FOR FOOD, AND THAT IS IT WAS PLEASANT TO THE EYES**

**GENESIS 3:6A**

**WE USUALLY GET TO SEE THE FOOD BEFORE WE TASTE IT. THE EYES HAVE A POWERFUL ROLE IN WHAT YOU EAT. IF THE APPEARANCE IS NOT PLEASANT, WE AVOID THAT FOOD.**

**OUR EYES ARE THE SENSE ORGANS WE USED MOST WHEN WE EAT SOMETHING.**

Adam and Eve used their sensory organs to select and eat food: Their eyes, nose, tongue, skin (touch), the nervous and the digestive systems.

The chemistry of food involves visual receptors that send messages to different parts of the brain. The brain interprets these sensations.

Color is our first visual cue to get our attention. Different colors can elicit different emotions towards our food. Red and yellow are the most reflective colors that cause our taste buds to become activated and stimulate the appetite.

The Bible does not give any clues to what colors or taste the fruits were in the garden of Eden. However, it was perfectly delicious and nutritious.

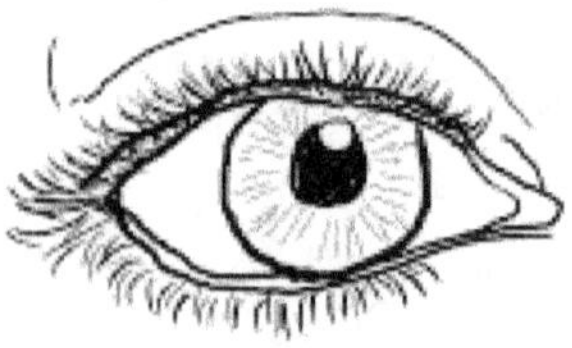

**THE BIBLE IS LARGELY SILENT ON THE EYE COLOR AND SKIN COLOR OF ADAM AND EVE. THEIR EYE COLOR IS BASED ON THE AMOUNT OF MELANIN PRESENT IN THE IRIS. SHADES OF EYE COLOR IS DETERMINED BY GENETIC OR GENES CONTRIBUTION OF BOTH PARENTS. ADAM AND EVE COULD HAVE CARRIED THE GENES FOR BLUE EYES TO BLACK EYES.**

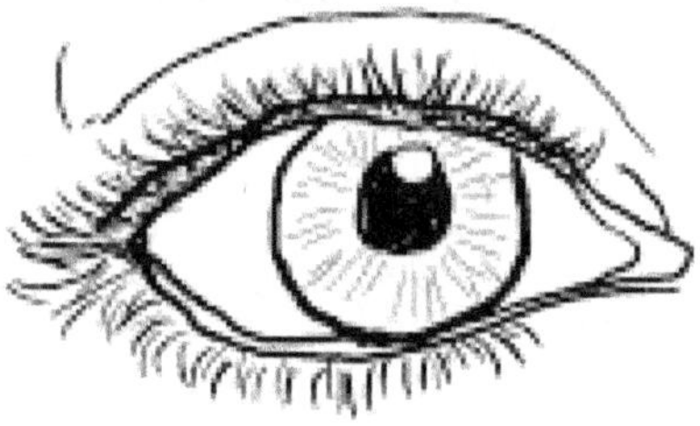

Gray

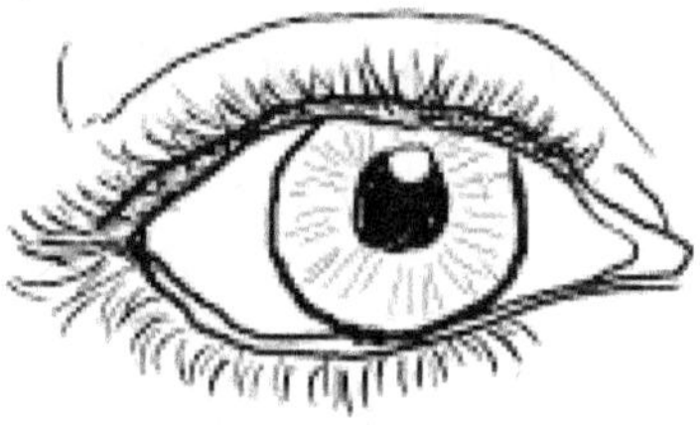

Hazel

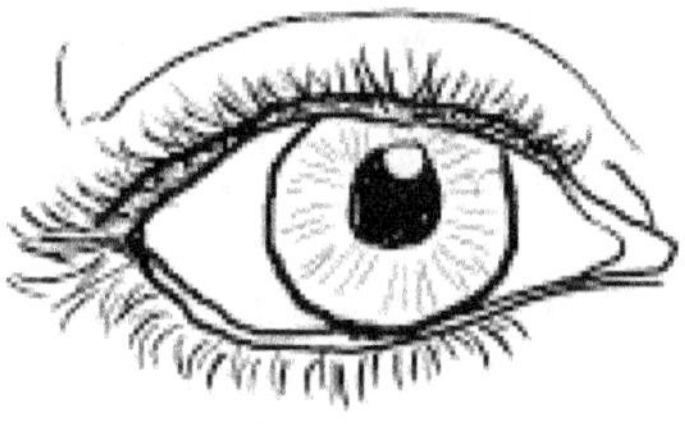

Green

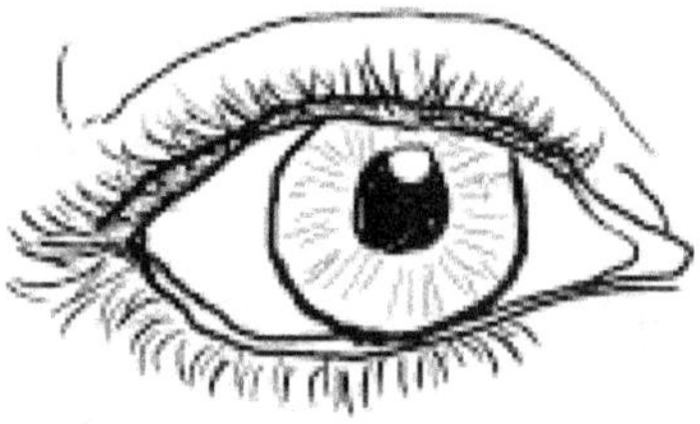

Brown

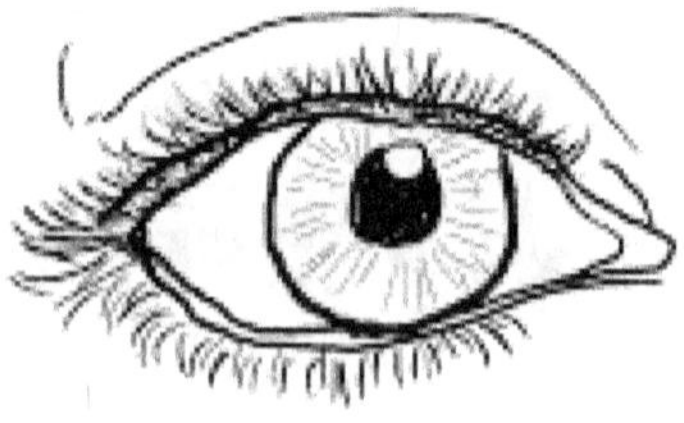

Blue

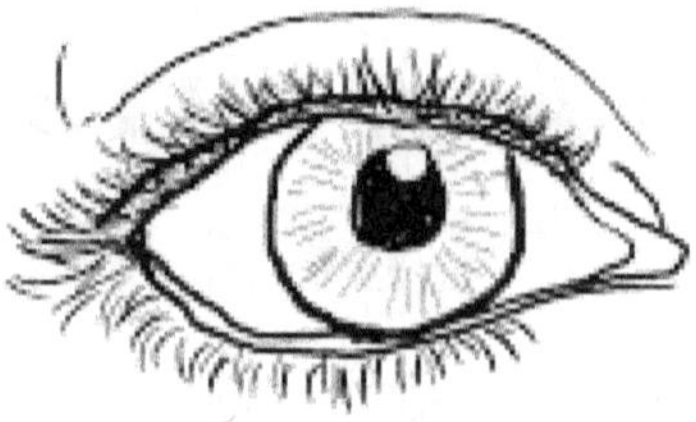

Violet

# Properties of Light - Reflection, Refraction, Absorption

"I have given every green herb for food"
Genesis 1:30b

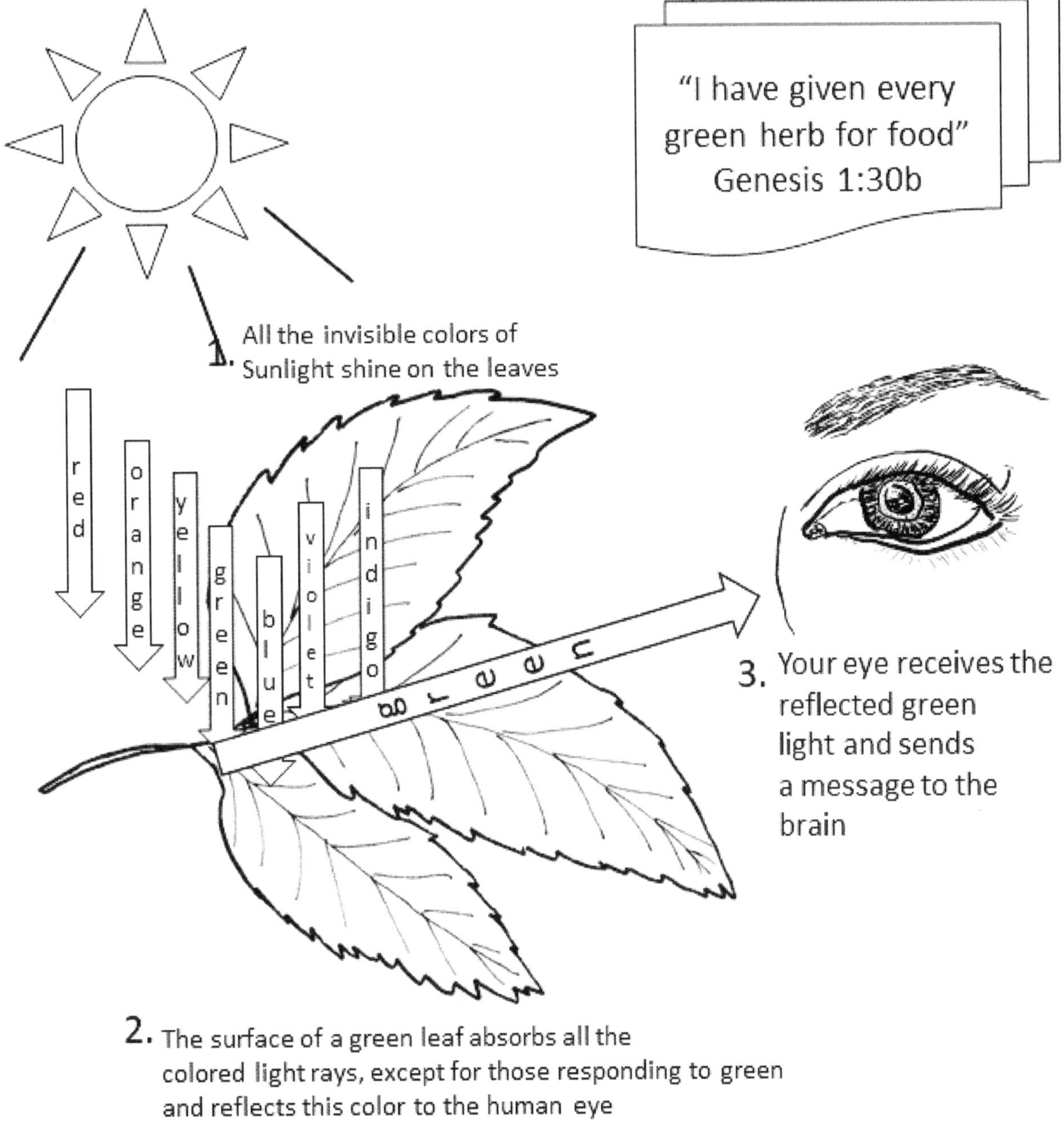

# Journal Notes

# Taste Flavors

Taste buds are found on your tongue and allow you to experience sweet, sour,

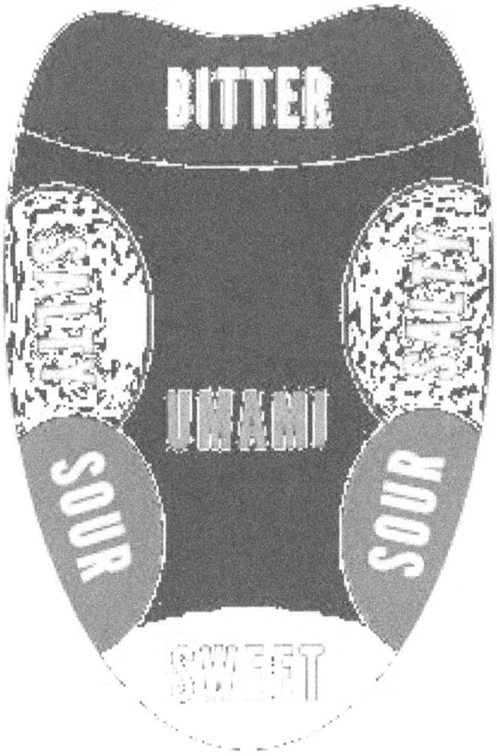


salty, sweet, bitter, umami. Your

tongue has about 2,000-10,00 taste buds that are replace every two weeks or so.

To taste food, it must be chewed and mixed with saliva. The taste message is sent to the brain by nerves.

Each of the taste sensations have special center on the tongue.

# Cornucopia - Garden of Eden

## Genesis 1: 29-30

**AND GOD SAID, BEHOLD I HAVE GIVEN YOU EVERY HERB BEARING SEED WHICH IS UPON THE FACE OF ALL THE EARTH AND EVERY TREE, IN WHICH IS THE FRUIT OF A TREE YIELDING SEED; TO YOU IT SHALL BE FOR MEAT. AND TO EVERY BEAST OF THE EARTH AND TO EVERY FOWL OF THE AIR, AND TO EVERYTHING THAT CREEPS UPON THE EARTH, WHEREIN THERE IS LIFE, I HAVE GIVEN EVERY GREEN HERB FOR MEAT: AND IT WAS SO.**

**Teach the younger women-**
**Titus 2:3-5**

In the beginning the rich fertile land was commanded by God to bring forth three types of vegetation:

the grass, herb

yielding seeds, and fruit of a tree- yielding seeds trees (Genesis 1:11-12;29)

The fruit trees that grew in the garden were pleasant to

the eyes and good for food.

Fruits, vegetables, nuts, seeds, grains, and legumes were available to eat.

The human body need nutrients to function properly. The nutrients that are essential for humans are proteins, carbohydrates, lipids, minerals, vitamins, and water.

FRUITS PROVIDED HIGH CONTENT OF CARBOHYDRATES (STARCH AND SUGAR) , VITAMINS, MINERALS, AND FIBERS. CARBOHYDRATE PROVIDE THE MAIN SOURCE OF THE BODY'S ENERGY. SUGARS ARE NOT ESSENTIAL FOODS ALTHOUGH THEY PROVIDE ENERGY BUT NOT NUTRIENTS.

NUTS ALSO GROW ON TREES. NUTS CONTAIN MOST OILS OR FATS. ADAM AND EVE COULD EAT RAW NUTS LIKE ALMONDS (GENESIS 43:11), WALNUTS, PECANS, BRAZIL NUTS, PISTACHIO NUTS (NUMBERS 18:7), AND FILBERTS. NUTS ARE A GOOD SOURCE OF PROTEINS, VITAMIN E, MINERALS SUCH AS MAGNESIUM, PHOSPHORUS, IRON, AND FIBERS. PROTEINS ARE NECESSARY FOR GROWTH AND TISSUE REPAIR.

FATS AND OILS ARE A CONCENTRATED SOURCE OF ENERGY. THEY PROVIDE VITAMINS FOR THE BODY CUSHION VITAL ORGANS & CELLS AND MAINTAIN BODY TEMPERATURE.

Vine bearing vegetables and fruits with edible seeds were most likely available. They could be cucumbers, tomatoes, peppers, berries, grapes, zucchini, and small melons.

There is no recorded event in the Bible of the use of fire in the Garden of Eden. So, we can assume that Eve did not cook their meals. Food was easy to obtain in the garden and was probably picked and eaten immediately. Adam was given the responsibility to cultivate the garden. He knew where to find the best food source.

# Cheiloscopy - study of Lip Patterns

## Psalm 19:14

LET THE WORDS OF MY MOUTH .. .BE ACCEPTABLE IN YOUR SIGHT

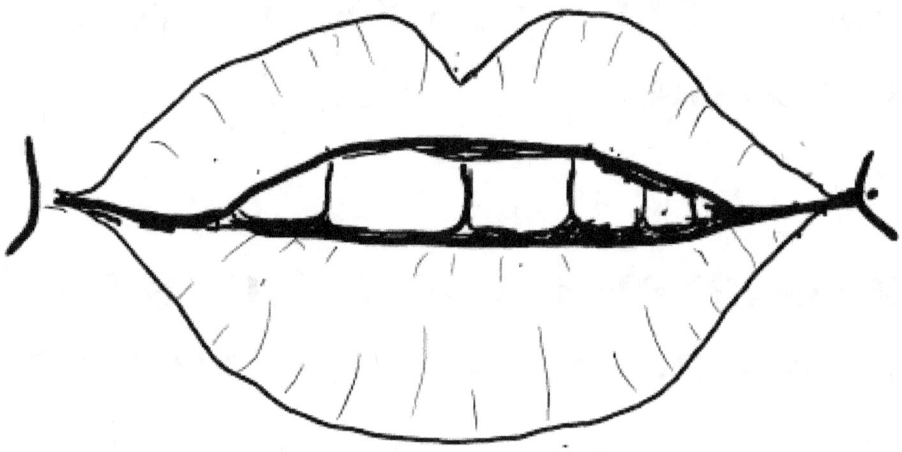

GOD CREATED OUR BODY WITH UNIQUENESS. LIP PRINTS WERE DESIGNED IN THE WOMB AND ARE PERMANENT. LIPS ARE SOFT, MOVABLE, SERVE AS THE OPENING FOR FOOD INTAKE, AND AID IN THE ARTICULATION OF SOUND AND SPEECH. THE SKIN OF THE LIPS IS VERY THIN AND CONTAIN BLOOD VESSELS. BECAUSE OF NO SWEAT GLAND THE LIPS CAN GET VERY DRY AND BECOME CHAPPED.

**FUNCTIONS OF THE LIPS**

- FOOD INTAKE – MUSCLES ALLOW THE LIP TO BE VERY MOVABLE FOR EATING, SUCKING, CLOSING THE MOUTH, AND CREATING A VACUUM SEAL SO WE CAN SWALLOW.
- ARTICULATION- LIPS SERVE FOR CREATING DIFFERENT SOUNDS LIKE WHISTLING, READING LIPS WITHOUT SPEECH
- TACTILE ORGAN- THE LIPS ARE SENSITIVE AND HAS MANY NERVE ENDINGS THAT CAUSES HUMANS TO BE MORE SENSITIVE TO WARMTH, COLD TEMPERATURE, AND KISSING

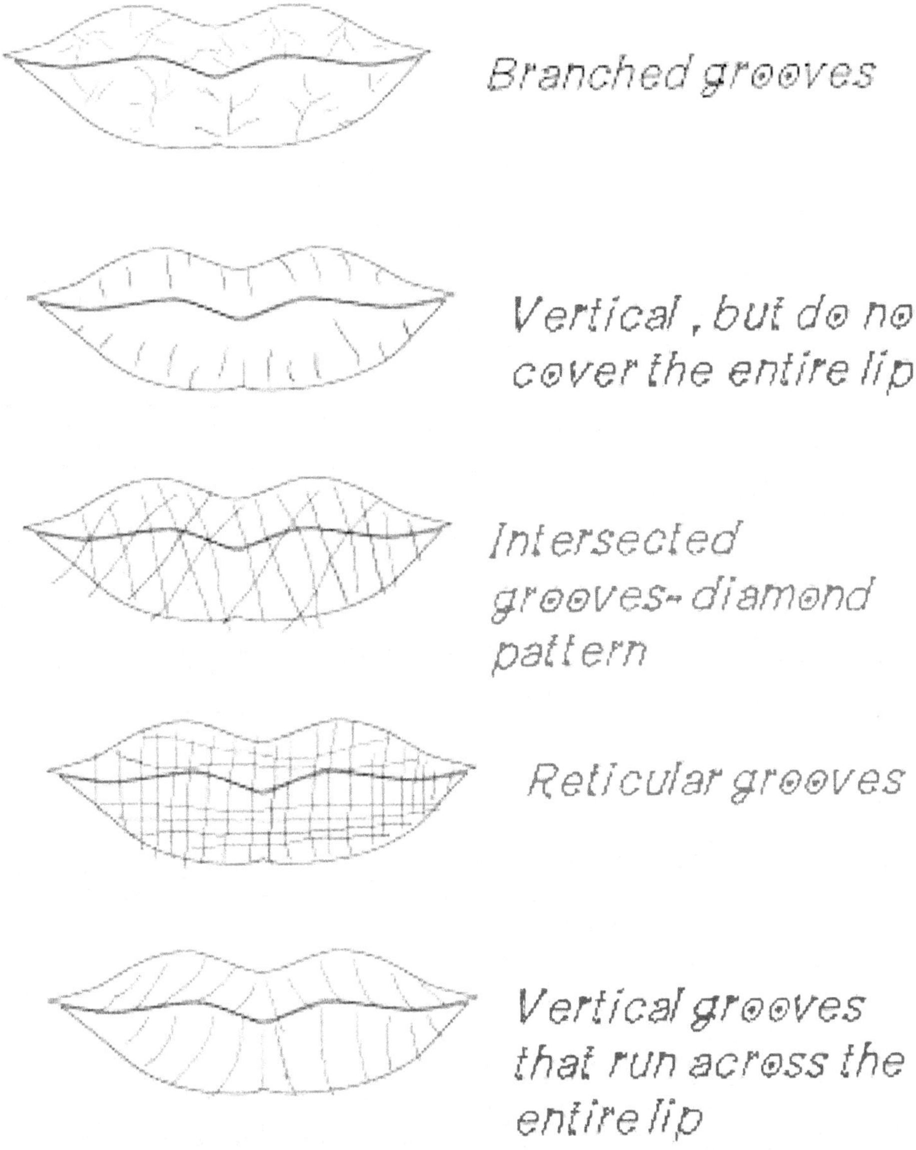
Branched grooves
Vertical , but do not cover the entire lip
Intersected grooves- diamond pattern
Reticular grooves
Vertical grooves that run across the entire lip

# Olfactory - Nose

THE NOSE FUNCTIONS TO CLEAN THE AIR PASSING THROUGH FINE HAIRS AND MUCOUS TO REMOVE DIRT AND DUST PARTICLES BEFORE ENTERING THE LUNGS. THE NASAL CAVITIES ARE ALSO LINED WITH TINY BLOOD VESSELS THAT MOISTEN, AND WARMS AIR DRAWN IN BY THE LUNGS. OUR NOSE ASSISTS US IN SMELLING OUR FOOD. ALSO, AIR RESONATING IN YOUR NOSE AFFECTS THE SOUND OF YOUR VOICE.

THE NOSE IS THE FIRST ORGAN OF THE RESPIRATORY SYSTEM. A HUMIDIFIER IS USED TO INCREASE THE AMOUNT OF WATER VAPOR IN THE AIR. YOUR NOSE HUMIDIFIES THE AIR AND ADD MOISTURE TO THE AIR BEFORE IT GETS INTO YOUR LUNGS.

WHEN YOU ARE CHEWING YOUR FOOD RELEASE FLAVOR CHEMICALS THAT TRAVEL UP INTO YOUR NOSE. THESE STIMULATIONS WILL TRIGGER THE OLFACTORY RECEPTORS INSIDE THE NOSE.

YOUR NOSE AND YOUR TASTE BUDS WORK TOGETHER TO CREATE THE FLAVOR THAT TELLS YOUR BRAIN "THAT'S YUMMY."

CONDITIONS SUCH AS A COLD WILL DECREASE YOUR SENSE OF SMELL AND THE TASTE OF YOUR FOOD.

# Fingerprint Patterns

***Psalms 139:13 – 16***

*"I WILL PRAISE YOU, FOR I AM FEARFULLY AND WONDERFULLY MADE; MARVELOUS ARE YOUR WORKS. AND THAT MY SOUL KNOWS VERY WELL. MY FRAME WAS NOT HIDDEN FROM YOU, WHEN I WAS MADE IN SECRET AND SKILLFULLY WROUGHT IN THE LOWEST PARTS OF THE EARTH. YOUR EYES SAW MY SUBSTANCE, BEING YET UNFORMED, AND IN YOUR BOOK THEY ALL WERE WRITTEN, THE DAY FASHION FOR ME, WHEN YET THERE ARE NONE WRITTEN."*

**ALTHOUGH EVE WAS NOT BORN BUT CREATED HER FIRST-BORN CHILD HAD THE MIRACLE OF THE WOMB.**

**CAIN'S RIDGES ON HIS FINGERS STARTED TO DEVELOP WHILE IN HIS MOTHER'S, EVE, WOMB. A FETUS WILL BEGIN DEVELOPING ITS OWN UNIQUE SET OF FINGERPRINTS BY THE 3-MONTH GESTATION MARK. IT IS BELIEVED THAT THE ENVIRONMENT OF THE WOMB HAS A DIRECT INFLUENCE ON THE DEVELOPMENT OF FINGERPRINTS.**

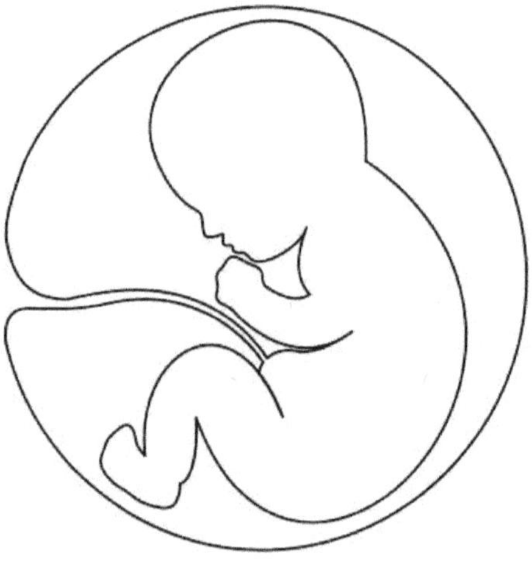

Be steadfast,
unmovable –
1 Corinthians
15:58

# Journal Notes

# Dactyloscopy is the practice of using fingerprints to identify someone

**THE PATTERNS OR IMPRESSIONS ON OUR FINGERS ARE UNIQUE. NO TWO INDIVIDUALS, EVEN IDENTICAL TWINS, HAVE FINGERPRINTS THAT ARE EXACTLY ALIKE. FINGERPRINTS CAN BE VISIBLE WHEN OUR FINGERS ARE DIRTY.**

**FINGERPRINTS CAN BE CLASSIFIED BY PATTERNS TIGHTS, BY THE SIZE OF THESE PATTERNS. WE LEAVE IMPRESSION PRINTS OF THESE PATTERNS ON EVERYTHING WE TOUCH WITH ANY PRESSURE.**

Loops

Whorls

Arches

# Your Hairs are Numbered

## Luke 12:7a

**THE FIRST MENTION OF HAIR IN THE BIBLE IS ABOUT ESAU, THE BROTHER OF JACOB BEING A HAIRY MAN (GENESIS 25:25)**

THE BIBLE DOES NOT GIVE INSIGHT INTO THE COLOR OR TEXTURE OF ADAM OR EVE'S HAIR.

THE WORLD'S POPULATION REVEALS BLACK, BROWN, BLONDE, AND RED ARE THE MAJOR HAIR COLORS.

HAIR IS A PROTEIN FILAMENT THAT GROWS FROM FOLLICLES FOUND IN THE DERMIS. HAIR COVER THE ENTER SURFACE OF THE SKIN.

HAIR IS VISIBLE ON THE SCALP, EYELIDS, AND EYEBROWS. NO HAIR FOLLICLES ARE FOUND ON THE SIDES AND SOLES OF THE FEET, THE PALMS OF THE HANDS, SIDES OF THE FINGER AND TOES, AND THE LIPS.

HAIR COLOR REFLECTS DIFFERENCES IN THE TYPE AND AMOUNT OF PIGMENT PRODUCED BY MELANIN CELLED MELANOCYTE IN THE HAIR FOLLICLES.

MELANIN GIVES US DARK-BROWN, YELLOW-BROWN, OR RED COLORATION TO THE HAIR.

## Hair textures includes:

- Straight - is the strongest type and reflects light to the eye giving the hair a glossy appearance

- Wavy- this type has a S-shaped curls

- Curly - S-shaped or corkscrew curls usually voluminous

- Kinky-curly-tight S-shaped or Z-shaped curls

Skin is the biggest organ of the body. The skin serves as a barrier to prevent injury to organs, repel dirt and water, absorbs ultraviolet radiation, Vitamin D production contains sensory receptors that can detect temperatures, and block harmful toxins. The skin also contains sweat gland that cool the body, lubricate with oil, secretion to kill harmful bacteria and prevents dehydration.

The original couple, Adam and Eve, pigmentation is not clearly stated. Adam and Eve were neither white nor black. Their skin color was most likely reddish tan or copper tone. The name Adam had Hebrew origin meaning "son of the red earth". This refers to the ruddy color of human skin. Adam was created from the dust of the earth or clay (Genesis 2:7).

# The Skin

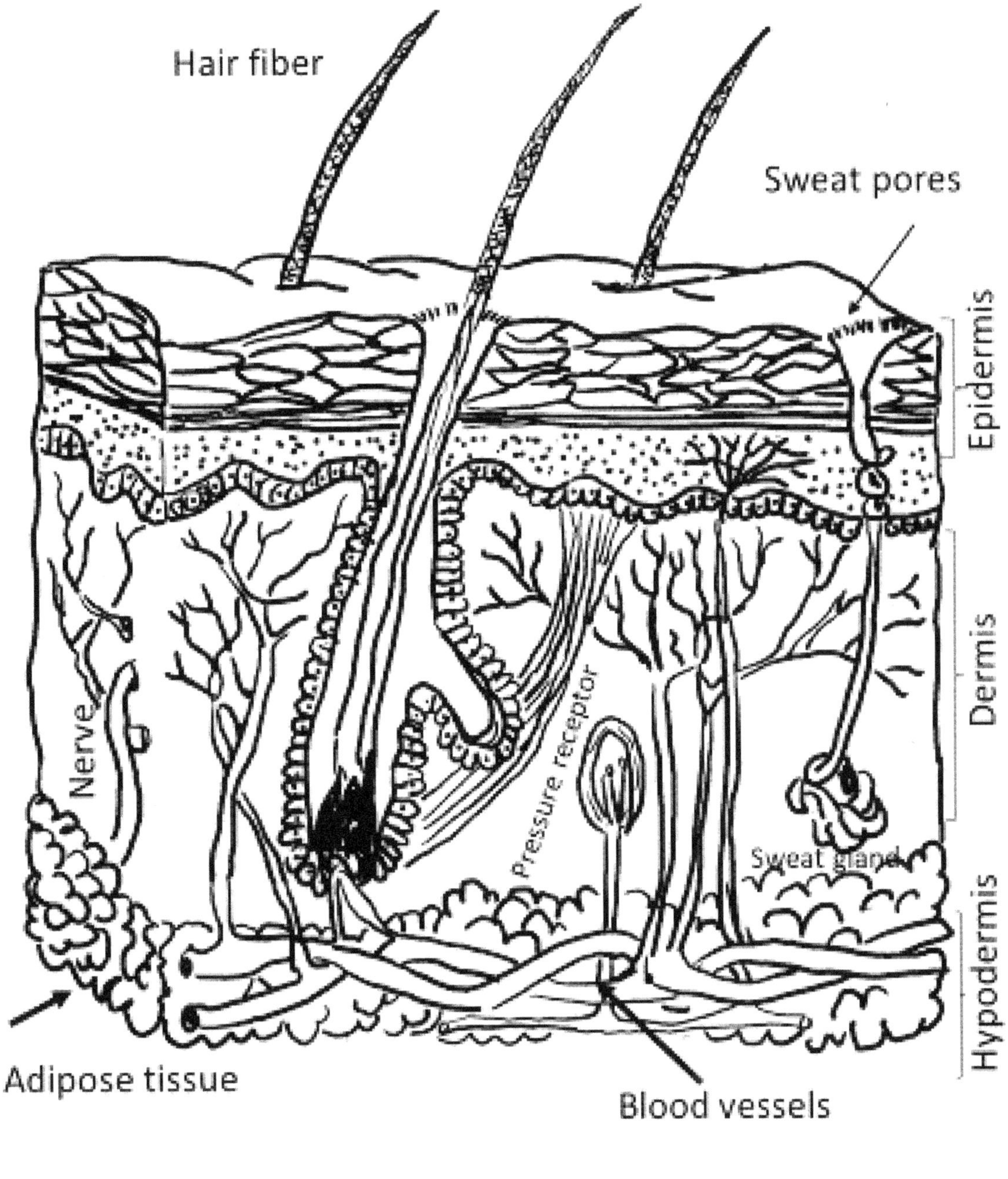

Hair fiber
Sweat pores
Epidermis
Dermis
Hypodermis
Nerve
Pressure receptor
Sweat gland
Adipose tissue
Blood vessels

# Human Skin Tones

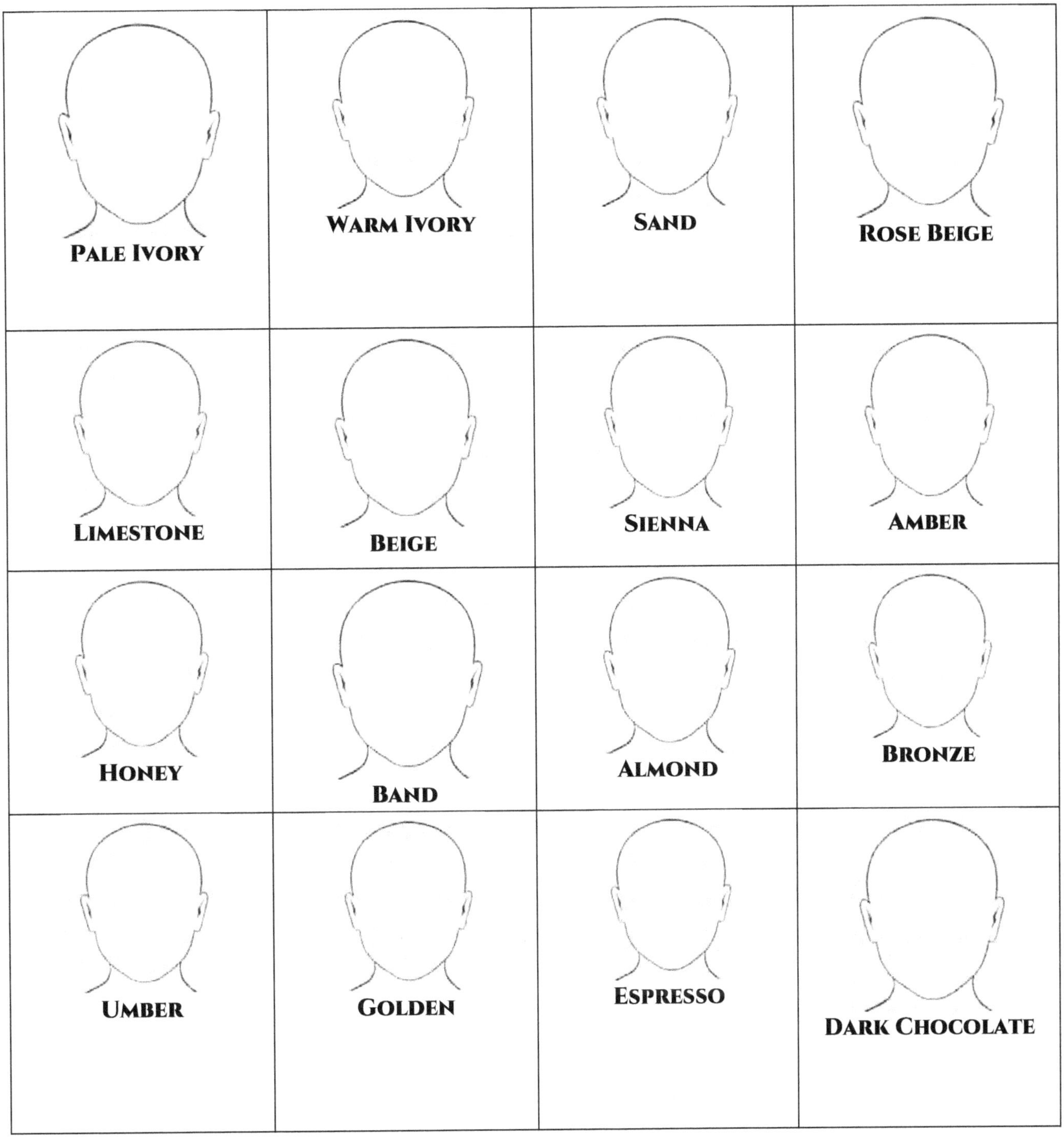

IF ADAM OF EVE WERE GENETICALLY MEDIUM BROWN, THEY WOULD PRODUCE CHILDREN WITH A WIDE RANGE OF TONES.

ALL PEOPLE HAVE MELANIN. MELANIN CELLS CALLED MELANOCYTES PROVIDE COLORING TO THE ENTIRE EPIDERMIS, EYE COLOR AND HAIR COLOR. MELANIN IS A BROWN, YELLOW-BROWN OR BLACK PIGMENT.

MELANOCYTE RESPONSE TO SUNLIGHT THAT CONTAINS ULTRAVIOLET RADIATION. MELANIN HELPS PREVENT SKIN DAMAGE BY ABSORBING ULTRAVIOLET RADIATION BEFORE IT REACHES THE DEEP LAYERS OF THE SKIN. MELANIN SURROUNDS THE NUCLEUS AND HELPS TO BUILD A RADIATION SHIED OF PROTECTION OF THE NUCLEAR DNA. TOO MUCH EXPOSURE COULD CAUSE SKIN CANCER.

# The Woman- created for Purpose and Function

THE MEANING OF "HELP MEET" STEMS FROM THE HEBREW TERMS *EZER* AND *K'ENEDGO.* THESE HEBREW TERMS HAVE THEIR MEANING AS "TO RESCUE, TO SAVE, TO BE STRONG, AND SALVATION." THE STRONG'S HEBREW STATES EZER IS "TO SURROUND PROTECT, AID, OR SUCCOR." MERIAM-WEBSTER DICTIONARY STATES THE MEANING OF THE WORD SUCCOR IS SOMEONE THAT GIVES ASSISTANCE AND SUPPORT IN TIMES OF HARDSHIP AND DISTRESS. A DISTINCT KIND OF HELP.

# Journal Notes

# Priscilla and Aquila

## Acts 18:1-26

# Journal Notes

# Worship

THE GARDEN OF EDEN AS THE FIRST PLACE ON EARTH FOR WORSHIP

# Journal Notes

God placed the new couple in the Garden. They have an open invitation to be with Him in personal relationship with the Creator. God was looking at their heart. God provided a place for them to live, good food to eat, dominion on the Earth. God only request was to acknowledge and obey Him.

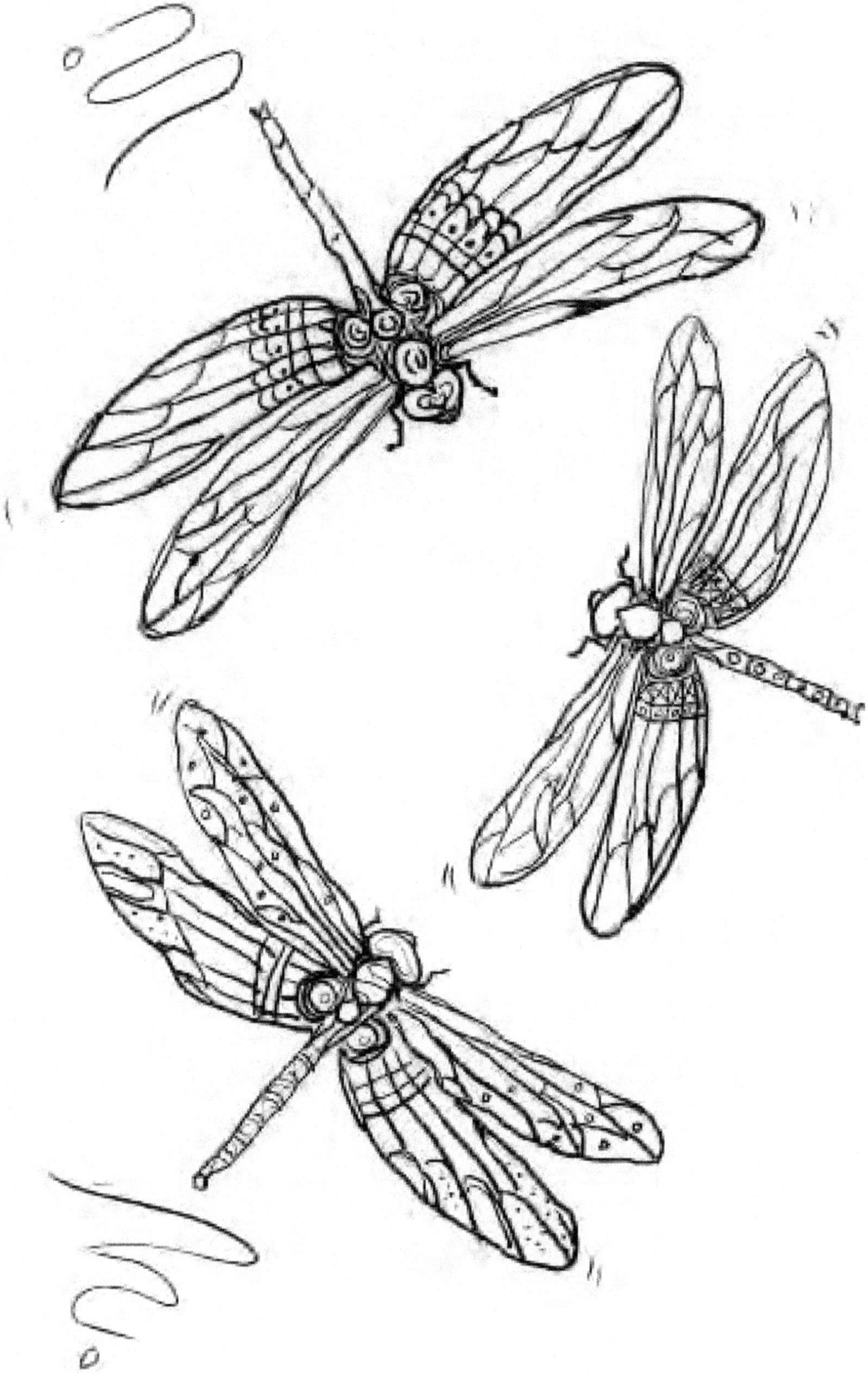

# The Family Begins

Genesis 4:1

To Be Continued . . . . . . .

www.ingramcontent.com/pod-product-compliance
Lightning Source LLC
LaVergne TN
LVHW060450061126
851857LV00016B/1779